R. P. BURN

GROUPS

A PATH TO GEOMETRY

The right of the
University of Cambridge
to print and sell
all manner of books
was granted by
Henry VIII in 1534.
The University has printed
and published continuously
since 1584.

CAMBRIDGE UNIVERSITY PRESS

Cambridge

London New York New Rochelle

Melbourne Sydney

Published by the Press Syndicate of the University of Cambridge
The Pitt Building, Trumpington Street, Cambridge CB2 1RP
32 East 57th Street, New York, NY 10022, USA
10 Stamford Road, Oakleigh, Melbourne 3166, Australia

First published 1985

Printed in Great Britain at the University Press, Cambridge

Library of Congress catalogue card number: 84-21354

British Library cataloguing in publication data
Burn, R. P.
Groups: a path to geometry.
1. Groups, Theory of
I. Title
512'.2   QA171

ISBN 0 521 30037 1

Contents

tributive laws, the set is called a field when the multiplicative group is commutative. When a multiple direct product is formed with the same additive group of a field as each component, and this direct product is supplied with a scalar multiplication from the field, the direct product is called a vector space.

When two vector spaces have the same field, a structure preserving function of one to the other can be described with a matrix.

The structure preserving permutations of a 2-dimensional vector space are analysed.

Scalar and vector products are defined in three dimensions. The meaning and properties of determinants of 3×3 matrices are explored.

Vectors mapped onto scalar multiples of themselves under a linear transformation are found and used to construct a diagonal matrix to describe the same transformation, where possible.

Those functions of a group to a group which preserve the multiplicative structure are analysed. The subset of elements mapped to the identity is a normal subgroup. Each coset of that normal subgroup has a singleton image.

When x and g belong to the same group, the elements x and $g^{-1}xg$ are said to be conjugate. Conjugate permutations have the same cycle structure.

Contents

Groups of isometries not fixing a point or a line are shown to contain translations. If there are no arbitrarily short translations, the translation group has two generators. If such a group contains rotations, their order may only be 2, 3, 4 or 6. The possible point groups are then C_1, C_2, C_3, C_4, C_6, D_1, D_2, D_3, D_4 or D_6. This provides a basis for classifying the seventeen possible groups of this type.

Preface

This book contains a first course in group theory, pursued with conventional rigour. There are three unusual aspects of the presentation.

Firstly, the book consists of a sequence of over 800 problems. This is to enable the course to proceed by seminar rather than by lecture. Mathematics is something we do rather than something we learn, and, all too often, lectures give the opposite impression.

Secondly, at the outset, the groups under discussion are groups of transformations. This is faithful to the historical origins of the theory. It provides the one context in which the proof of the associative law is immediate, and it makes the study of sets with only a single defined operation obviously worthwhile. For Galois (1830), Jordan (1870) and even in Klein's '*Lectures on the Icosahedron*' (1884), groups were defined by the one axiom of closure. The other axioms were implicit in the context of their discussions – finite groups of transformations. Our work on abstract groups starts in chapter 6.

Thirdly, the geometry of two and three dimensions is the context in which most of the groups in this book are constructed, and is also the major field of application of group theory in chapters 7, and 17–23. Geometry is the best context in which to understand conjugacy, and linear and affine groups are some of the easiest in which to put homomorphisms to work to good effect. The geometrical bias of the book links group theory with complex analysis, linear algebra and crystallography and provides a useful background for anyone about to study these fields.

The mathematical knowledge assumed for this course is a confident familiarity with high school mathematics. In a few places an argument by induction is needed and in one place the cosine rule is required. Some experience at school level with groups, with matrices and with complex numbers, would certainly be helpful, though the text is self-contained in these respects. Although this book assumes less previous knowledge than my earlier *Pathway to Number Theory*, the sequence of questions on geometrical groups may cause more difficulty to the solitary student. The cumulative development of concepts is heavier in this area. Partly to compensate for this, ideas required from outside group theory are developed concretely and with the minimum of abstraction. For example, abstract vector spaces are not defined in this book: only spaces of n-tuples are used. It may be reassuring to the student to know that the results of chapters 7, 18 and 19 are not used in the rest of the book.

A great many of the results obtained in this book have been claimed and quoted by schoolteachers during the last 20 years, but many of the proofs have been inaccessible. I hope I have provided here, not just proofs, but also insight into transformation geometry as it is now done in British schools. For the undergraduate, there is a broad concrete base for generalisation and abstraction in his further studies.

Acknowledgements

I am very grateful to Dr Alan Beardon for partnership in designing the course for which this book is a text and to my colleague, Bob Hall, for many hours of discussion about the actual questions and their solutions. I am grateful too for correspondence on historical matters with Dr P. Neumann and Prof. B. L. van der Waerden. The faults that remain are my own.

Homerton College, Cambridge   R. P. Burn
July 1984

1

Functions

Throughout the nineteenth century, group theory was a study of permutations and substitutions. Group elements were generally referred to as 'operations', being what we would now call transformations (bijections) of a set to itself. This view of groups suits geometers very well, and it is the view that we adopt for most of this book.

In this first chapter we establish those properties of transformations which make sets of transformations form groups under composition and we do this with a set-theoretic rigour unknown in the nineteenth century. This chapter is the most abstract chapter in the book, and the student who finds this uncomfortable may start at chapter 2 provided that he accepts the results of the last question of chapter 1.

Because we will be establishing a formal definition of a function in this chapter we will also be providing a background for the terms isomorphism, homomorphism and one–one correspondence, all of which describe special kinds of functions which are of use in group theory, which are not usually thought of as group elements themselves.

Concurrent reading: Green, chapter 3.

FUNCTIONS $N \to L$

Arrow diagram   *Graph*

N = numbers   *L = letters*

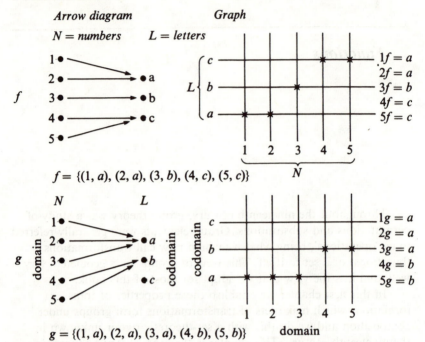

$$1f = a$$
$$2f = a$$
$$3f = b$$
$$4f = c$$
$$5f = c$$

$$f = \{(1, a), (2, a), (3, b), (4, c), (5, c)\}$$

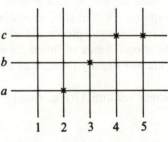

$$1g = a$$
$$2g = a$$
$$3g = a$$
$$4g = b$$
$$5g = b$$

$$g = \{(1, a), (2, a), (3, a), (4, b), (5, b)\}$$ domain

Domain $N = \{1, 2, 3, 4, 5\}$, Codomain $L = \{a, b, c\}$.

Both f and g are examples of functions from N to L,

$$f : N \to L \text{ and } g : N \to L$$

NOT a FUNCTION $N \to L$

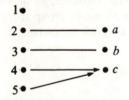

NOT a FUNCTION $N \to L$

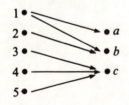

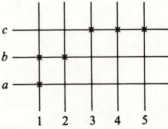

1 Use the diagrams on page 2, with their implied rules to complete the following sentence.

A function $f: N \to L$ is defined when for each element n of the set N there is. . . . .

2 Use the diagrams on page 2, with their implied rules to complete the last sentence.

The rectangular array used for the graphs of the functions on page 2 represents the so-called *cartesian product* $N \times L$. Each element of $N \times L$ is an ordered pair (n, x) with n from N and x from L. The graph of a function $N \to L$ consists of a subset of $N \times L$ with exactly one element (n, x) for each . . . .

3 Which of these define functions of the real numbers $\mathbb{R} \to \mathbb{R}$?

   (i) $x \mapsto x^2$,

   (ii) $x^2 \mapsto x$,

   (iii) $x \mapsto 1/x$,

   (iv) $x \mapsto \sin x$,

   (v) $x \mapsto \tan x$.

4 If $A = \{0, 1\}$, how many different functions, $A \to A$, are there?

Injections (one–one)

5 Draw sketch graphs of the functions $\mathbb{R} \to \mathbb{R}$ given by

(i) $x \mapsto x^3$ and by (ii) $x \mapsto x^2$.

If $x^3 = y^3$, does it follow that $x = y$?
If $x^2 = y^2$, does it follow that $x = y$?

    The distinction here leads us to call $\alpha : \mathbb{R} \to \mathbb{R}$ defined by $\alpha : x \mapsto x^3$, a *one–one* function or *injection*. We say that $\beta : x \mapsto x^2$ is *not* one–one on $\mathbb{R}$.

6 Let $A = \{1, 2, 3, 4\}$.

   (i) Exhibit the graph of a one–one function $A \to A$.

   (ii) Exhibit the graph of a function $A \to A$ which is not one–one.

   (iii) How many functions $A \to A$ exist?

7 Let $\mathbb{N}$ denote the set of natural numbers $\{1, 2, 3, \ldots\}$. Draw part of the graph of the function $\mathbb{N} \to \mathbb{N}$ defined by $n \mapsto n^2$. Is this a one–one function?

8 What can be said about the rows of the graph of a function if that function is known to be one–one?

Surjections (onto)

9 Draw sketch graphs of the functions $\mathbb{R} \to \mathbb{R}$ given by

(i) $\alpha : x \mapsto e^x$ and (ii) $\beta : x \mapsto x + 1$.

Can you always find a real number x, such that $\alpha : x \mapsto y$ for any choice of the real number y?

Can you always find a real number x, such that $\beta : x \mapsto y$ for any choice of the real number y?

The distinction here leads us to call β an *onto* function or *surjection* and to say that α is *not* onto.

10 Let $A = \{1, 2, 3, 4\}$. Exhibit the graph of a function $A \to A$
  (i) which is onto,
  (ii) which is not onto.

11 Illustrate part of the graph of the function $\mathbb{N} \to \mathbb{N}$ defined by

the even number $2n \mapsto n$,
the odd number $2n - 1 \mapsto n$.

Is this a one–one function?
Is this an onto function?

12 What can be said about the rows of the graph of an onto function?

13 Let $A = \{1, 2, 3, 4\}$.

Can you construct a function $A \to A$ which is one–one but not onto?

Can you construct a function $A \to A$ which is onto but not one–one?

14 Can you construct a function $\mathbb{N} \to \mathbb{N}$ which is one–one but not onto?

Can you construct a function $\mathbb{N} \to \mathbb{N}$ which is onto but not one–one?

15 Conjecture a condition on an arbitrary set A such that every one–one function $A \to A$ must be onto, and every onto function $A \to A$ must be one–one. Justify your conjecture with the help of qn 8 and qn 12.

An injection which is also a surjection is called a *bijection*. In the context of functions with the same finite set as domain and codomain, injections, surjections and bijections are in fact indistinguishable.

16 Give examples of functions $\mathbb{R} \to \mathbb{R}$ which are
  (i) one–one and onto (bijections),
  (ii) one–one but not onto (injections),
  (iii) onto but not one–one (surjections),
  (iv) neither one–one nor onto.

17 If there exists a one–one function $A \to A$ which is not onto, what can be said about the set A?

Composition of functions

18 If α and β are functions $\mathbb{R} \to \mathbb{R}$ defined by

$\alpha : x \mapsto 2x$ and $\beta : x \mapsto x + 1$,

then we *define* $\alpha\beta : x \mapsto 2x + 1$.

$$x \overset{\alpha}{\mapsto} 2x \overset{\beta}{\mapsto} 2x + 1.$$

We write this $(x)\alpha\beta = (2x)\beta = 2x + 1$.
Determine $(x)\beta\alpha$ under a similar definition.

19 If $\alpha : A \to B$ and $\beta : B \to C$ are functions, give a formal definition of $\alpha\beta : A \to C$ by determining $(x)\alpha\beta$ (the image of x under the function $\alpha\beta$: first α, then β).

The function $\alpha\beta$ is called the *composite* of the function α and β.

20 If and

exhibit $\alpha\beta$.

21 What formal condition makes $\alpha : A \to B$ a one–one function or injection?

22 If $\alpha : A \to B$ and $\beta : B \to C$ are injections, what can be said about $\alpha\beta : A \to C$?

23 Let $A = \{1, 2, 3\}$ and $B = \{1, 2, 3, 4\}$ and let $\alpha : A \to B$ be an injection.
Can you construct a function $\beta : B \to A$ such that
(i) $\beta\alpha$ is the identity function $B \to B$,
(ii) $\alpha\beta$ is the identity function $A \to A$?
(Under the *identity function* on the set, each point is its own image.)

Inverse functions

24 Let A and B be arbitrary sets, with $\alpha : A \to B$ an injection.
Show how to define $\beta : B \to A$ such that $\alpha\beta$ is the identity function on A.

The function β is then called a *right inverse* for α. So injections have right inverses.

25 What formal condition makes $\alpha: A \to B$ an onto function or surjection?

26 If $\alpha: A \to B$ and $\beta: B \to C$ are surjections, what can be said about the composite function $\alpha\beta: A \to C$?

27 Let $A = \{1, 2, 3, 4\}$ and $B = \{1, 2, 3\}$ and let $\alpha: A \to B$ be a surjection.
  (i) Can you construct a function $\beta: B \to A$ such that $\alpha\beta$ is the identity function on A?
  (ii) Can you construct a function $\beta: B \to A$ such that $\beta\alpha: B \to B$ is the identity function on B?

28 Let A and B be arbitrary sets, with $\alpha: A \to B$ a surjection. Show how to define a function $\beta: B \to A$ such that $\beta\alpha$ is the identity function on B.

The function β is then called a *left inverse* for α. So surjections have left inverses.

29 Let $A = \{1, 2, 3\}$. Make a list of the bijections $A \to A$, or in other words of the permutations of A.
Find a left inverse for each bijection.
Find a right inverse for each bijection.

30 For any bijection $\alpha: A \to B$, define a bijection $\beta: B \to A$ such that $\alpha\beta$ is the identity function $I: A \to A$ and $\beta\alpha$ is the identity function $B \to B$.
Prove that *either* $\alpha\beta = I: A \to A$ *or* $\beta\alpha = I: B \to B$, determines β uniquely.
So bijections have two-sided inverses.

Closure

31 If α and β are both bijections $A \to A$, what can be said about $\alpha\beta: A \to A$?

Associativity

32 Let $\alpha: A \to B$, $\beta: B \to C$ and $\gamma: C \to D$ be functions. Use your definition of qn 19 to show that for each point x of A,

$$(x)\,[(\alpha\beta)\gamma] \;=\; (x)\,[\alpha(\beta\gamma)],$$

so that the two functions $A \to D$, $(\alpha\beta)\gamma$ and $\alpha(\beta\gamma)$ are indistinguishable.

The theorem $(\alpha\beta)\gamma = \alpha(\beta\gamma)$ is called the *associative law* for functions under composition.

33 Use the associative law to prove for four functions $\alpha : A \to B$, $\beta : B \to C$, $\gamma : C \to D$, $\delta : D \to E$ that $\alpha[\beta(\gamma\delta)] = [(\alpha\beta)\gamma]\delta$.

The symmetric group

34 Let S be the set of all bijections $A \to A$.
Justify each of the following claims.
  (i) If α and β are in S, then $\alpha\beta$ is in S (*closure*).
 (ii) If α, β and γ are in S, then $\alpha(\beta\gamma) = (\alpha\beta)\gamma$ (*associativity*).
(iii) $I : A \to A$, defined by $I : x \mapsto x$ for all x in A, is in S (*identity*).
     $I\alpha = \alpha I = \alpha$ for all α in S.
 (iv) For each α in S there is a β in S such that $\alpha\beta = \beta\alpha = I$. (*inverses*)

These four theorems are summarised in the statement that the bijections of a set to itself form a *group* under composition. The group in this case is called the symmetric group on A and is denoted by S_A.

Summary

Definition qn 1	A *function* $\alpha : A \to B$, with domain A and codomain B is defined when, for each $a \in A$, there is a unique $b \in B$ such that $a\alpha = \alpha(a) = b$.
Definition qn 8	A function $\alpha : A \to B$ is an *injection* when, for any $a_1, a_2 \in A$, $a_1\alpha = a_2\alpha \Rightarrow a_1 = a_2$.
Definition qn 12	A function $\alpha : A \to B$ is a *surjection* when, for each $b \in B$, there exists an $a \in A$ such that $a\alpha = b$.
Definition qn 19	The *composite* (or *product*) $\alpha\beta$ of two functions $\alpha : A \to B$ and $\beta : B \to C$ is defined by $(a)\alpha\beta = (a\alpha)\beta$.
Theorem qn 22	The product of two composable injections is an injection.
Theorem qn 24	An injection has a right inverse.
Theorem qn 26	The product of two composable surjections is a surjection.
Theorem qn 28	A surjection has a left inverse.
Theorem qn 30	A bijection (an injection which is also a surjection) has a unique two-sided inverse.
Theorem qn 32	The composition of functions is associative.
Theorem qn 34	The set of bijections of a set to itself forms a group under composition. If the set is A, the group is called the *symmetric group* on A and is denoted by S_A.

Historical note

The modern notion of a function, with domain, and codomain, is essentially that of P. G. L. Dirichlet (1837). The language and style in which functions are discussed today owes much to the corporate twentieth-century French mathematician N. Bourbaki.

Answers to chapter 1

1 A function $f: N \to L$ is defined when for each element $n \in N$ there is a unique $l \in L$ with $nf = l$.

2 Exactly one element (n, x) for each $n \in N$.

3 (i) yes,   (ii) 1 has no unique image,   (iii) 0 has no image,
    (iv) yes,   (v) $\frac{1}{2}\pi$ has no image.

4 Four.

5 $x^3 = y^3 \Rightarrow x = y.$ $x^2 = y^2 \Rightarrow x = \pm y.$

6 (i)   (ii)   (iii) 4^4.

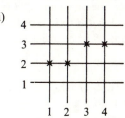

7 Yes.

8 Each row contains at most one entry.

9 (i) If $e^x = y$, y must be positive. (ii) If $x + 1 = y$, every y is possible.

10 (i)   (ii)

11 Onto, but not one–one.

12 Each row contains at least one entry.

13 No, no.

14 Yes, in qn 7.
Yes, in qn 11.

15 A must be finite. In this case the number of entries in the graph is equal to the number of rows. The condition that each row contains at least one entry is then equivalent to the condition that each row contains at most one entry.

16 (i) $x \mapsto x + 1$,   (ii) $x \mapsto e^x$,   (iii) $x \mapsto x^3 - x$,   (iv) $x \mapsto \sin x$.

17 A is infinite.

18 $(x)\beta\alpha = 2x + 2$. The left to right convention which we adopt is preferred by many algebraists. It has geometrical advantages in matrix algebra.

19 We define $\alpha\beta : A \to C$ by $(x)\alpha\beta = (x\alpha)\beta$.

20

21 $x\alpha = y\alpha \Rightarrow x = y$.

22 $x\alpha\beta = y\alpha\beta \Rightarrow x\alpha = y\alpha$ since β is an injection, $\Rightarrow x = y$ since α is an injection. Thus $\alpha\beta$ is an injection.

23 (i) No, because there is one point of B which is not an image under α. (ii) Yes. 1α, 2α and 3α are well defined and distinct. Let b be the fourth element of B. Define $(a\alpha)\beta = a$ for $a = 1, 2, 3$ and define $b\beta = 1$.

24 For $a \in A$ define $(a\alpha)\beta = a$. For $b \in B, b \neq a\alpha$ for any a, define $b\beta = a_1 \in A$.

25 Given $b \in B$, there exists $a \in A$ such that $a\alpha = b$.

26 Given $c \in C$, there exists $b \in B$ such that $b\beta = c$ and there exists $a \in A$ such that $a\alpha = b$ so $a\alpha\beta = c$ and $\alpha\beta$ is a surjection.

27 (i) No, because one point of A is not an image under β.
(ii) For each $b \in B$ there is at least one $a \in A$ such that $a\alpha = b$. For each $b \in B$, define $b\beta$ to be one element $a \in A$ such that $a\alpha = b$.

28 As in the second part of qn 27.

29 Images of $(1, 2, 3)$ are $(1, 2, 3)$, $(2, 3, 1)$, $(3, 1, 2)$, $(1, 3, 2)$, $(3, 2, 1)$, $(2, 1, 3)$. First, fourth, fifth and sixth bijections are self-inverse. Second and third bijections are inverses of each other.

30 For each $b \in B$ there is a unique $a \in A$ such that $a\alpha = b$. Define $b\beta = a$. If $\alpha\beta = I$, then $a\alpha\beta = a$ so $(a\alpha)\beta = a$. If $\beta\alpha = I$, then $b\beta\alpha = b = a\alpha$ for a unique a. Now α is an injection so $b\beta = a$, and, as before, $a\alpha\beta = a$.

31 From qn 22 and qn 26, $\alpha\beta$ is a bijection.

32 $x[(\alpha\beta)\gamma] = [x(\alpha\beta)]\gamma = [(x\alpha)\beta]\gamma = (x\alpha)\,(\beta\gamma) = x[\alpha(\beta\gamma)]$.

33 $\alpha[\beta(\gamma\delta)] = (\alpha\beta)\,(\gamma\delta) = [(\alpha\beta)\gamma]\delta$. This kind of argument can be extended to any finite product, to show that its value is independent of the position of the brackets.

34 (i) from qn 31, (ii) from qn 32, (iii) obvious, (iv) from qn 30.

2

Permutations of a finite set

Concurrent reading: Fraleigh, sections 4 and 5.

1 Let $A = \{1, 2, 3\}$ and let $B = \{1, 2, 3, 4\}$. The bijection $\alpha:A \to A$ with arrow diagram can be denoted by $\begin{pmatrix} 1 & 2 & 3 \\ 2 & 1 & 3 \end{pmatrix}$, and the bijection β: $B \to B$ with arrow diagram

can be denoted by $\begin{pmatrix} 1 & 2 & 3 & 4 \\ 3 & 2 & 4 & 1 \end{pmatrix}$.

Write down the six elements of S_A using this notation. Also write down the six elements of S_B each of which maps 1 to 2 using this notation.

A bijection $A \to A$ (for any set A) is often called a *permutation* of A.

2 When $A = \{1, 2, 3, \ldots, n\}$, the group S_A is usually denoted by S_n. When $n = 3$, a geometrical description of the elements of S_3 is available if we label the vertices of an equilateral triangle with 1, 2 and 3. The permutation $\begin{pmatrix} 1 & 2 & 3 \\ 1 & 3 & 2 \end{pmatrix}$ then corresponds to a reflection symmetry of the triangle and the permutation $\begin{pmatrix} 1 & 2 & 3 \\ 2 & 3 & 1 \end{pmatrix}$ to a rota-

tional symmetry through 120°. Describe the geometrical analogues
of the other permutations in S_3.

Cycle notation

3 If $\alpha = \begin{pmatrix} 1 & 2 & 3 & 4 & 5 \\ 2 & 3 & 4 & 5 & 1 \end{pmatrix}$, write down the permutation $\alpha \cdot \alpha$ which we
denote by α^2, the permutation $\alpha^2 \cdot \alpha$ which we denote by α^3, the
permutation $\alpha^3 \cdot \alpha$ which we denote by α^4, the permutation $\alpha^4 \cdot \alpha$
which we denote by α^5 and the permutation $\alpha^5 \cdot \alpha$ which we denote
by α^6.

Suggest a simple description of α^5 and of α^6.

Write down the images of 1 under the permutations α, α^2, α^3, . . . ,
and thereby record the sequence 1, 1α, $1\alpha^2$, $1\alpha^3$, ...

Record also the sequence 2, 2α, $2\alpha^2$, $2\alpha^3$, . . . and the sequence 3,
3α, $3\alpha^2$, $3\alpha^3$, ...

Because each of these sequences is periodic and runs through the
five digits in the same order, the permutation α is usually represented by (12345), or equally well by (23451), or (34512) or
(45123), or (51234), in what is called *cycle* notation.

4 With $\alpha = (12345)$, write down the permutations α^2, α^3 and α^4 in cycle
notation.

Although this notation only describes a rather special kind of permutation, its suggestiveness and compactness are useful in more
general settings.

5 Find a, b, c, d and e when $(134)(25) = \begin{pmatrix} 1 & 2 & 3 & 4 & 5 \\ a & b & c & d & e \end{pmatrix}$.

6 Find a, b and c when $\begin{pmatrix} 1 & 2 & 3 & 4 & 5 \\ 5 & 1 & 4 & 3 & 2 \end{pmatrix} = (1ab)(3c)$.

7 For any permutation α, we define $\alpha = \alpha^1$, $\alpha \cdot \alpha = \alpha^2$, $\alpha^2 \cdot \alpha = \alpha^3$ and
for any positive integer n, $\alpha^{n+1} = \alpha^n \cdot \alpha$. By an induction on m we
can prove that $\alpha^{n+m} = \alpha^n \cdot \alpha^m$ and also prove that $(\alpha^n)^m = \alpha^{nm}$ for
all positive integers n and m. Now we define α^0 to be the identity
permutation, α^{-1} to be the (two-sided) inverse of α, and we define

$\alpha^{-n} = (\alpha^{-1})^n$ for positive integers n. With these additional defini-
tions, we can prove by induction that $\alpha^{n+m} = \alpha^n \cdot \alpha^m$ and $(\alpha^n)^m = \alpha^{nm}$
for any integers n and m. Can you prove $\alpha^n \cdot \alpha^m = \alpha^m \cdot \alpha^n$ directly
from associativity?

8 If $\alpha = \begin{pmatrix} 1 & 2 & 3 & 4 & 5 \\ 3 & 5 & 4 & 1 & 2 \end{pmatrix}$, exhibit the sequences

1, 1α, $1\alpha^2$, $1\alpha^3$, ...

2, 2α, $2\alpha^2$, $2\alpha^3$, ...

3, 3α, $3\alpha^2$, $3\alpha^3$, ...

4, 4α, $4\alpha^2$, $4\alpha^3$, ...

5, 5α, $5\alpha^2$, $5\alpha^3$, ...

and determine a, b, c, d, e, f, p, q, r if

$\alpha = (1ab)(2c) = (3de)(5f) = (2p)(1qr)$.

Because the two cycles (134) and (25) have no digit in common
they are said to be *disjoint*.

9 Express the permutations

(i) $\begin{pmatrix} 1 & 2 & 3 & 4 & 5 \\ 3 & 4 & 5 & 1 & 2 \end{pmatrix}$, (ii) $\begin{pmatrix} 1 & 2 & 3 & 4 & 5 & 6 & 7 \\ 7 & 6 & 1 & 2 & 3 & 4 & 5 \end{pmatrix}$,

(iii) $\begin{pmatrix} 1 & 2 & 3 & 4 & 5 & 6 & 7 & 8 \\ 6 & 2 & 5 & 7 & 8 & 1 & 3 & 4 \end{pmatrix}$,

as products of disjoint cycles.

10 Exhibit the six elements of S_3 in cycle notation.

When $1\alpha = 1$, the cycle (1) is often dropped from the expression
of the permutation α as a product of cycles.

11 Exhibit the 24 elements of S_4 in cycle notation.

12 If $\alpha = (123456)$, express the permutations α^2, α^3, α^4 and α^5 as products
of disjoint cycles.

13 Express the product (12)(13) as a single cycle. Express the product
(13)(12) as a single cycle. Does (12)(13) = (13)(12)?

14 Does (12)(34) = (34)(12)?
If $a_i \neq b_j$ for any i or j, can you be sure that
$(a_1a_2 \ldots a_r)(b_1b_2 \ldots b_s) = (b_1b_2 \ldots b_s)(a_1a_2 \ldots a_r)$?

The property we have obtained here is usually described by saying
that disjoint cycles *commute*.

A cycle $(a_1a_2 \ldots a_n)$ is called a cycle of length n.
Cycles of length 1 are often not written down.
Cycles of length 2 are called *transpositions*.
Cycles of length n are called *n-cycles*.

15 To show that every permutation in S_n can be written as a product of
disjoint cycles, we consider an arbitrary permutation $\alpha \in S_n$, and
try to find one of the cycles of which it is composed by looking at
the sequence

$$1, 1\alpha, 1\alpha^2, 1\alpha^3, 1\alpha^4, \ldots .$$

How do you know that the digits in this sequence cannot all be
different? Can you be sure that some repetition occurs in this
sequence by the term $1\alpha^n$ at the latest? Suppose $1\alpha^m$ is the first term
which is a repetition of a preceding term in the sequence, and that
in fact $1\alpha^m = 1\alpha^k$ with $m > k \geqslant 0$. Then $1\alpha^{m-1} = 1\alpha^{k-1}$ and this
contradicts $1\alpha^m$ being the first repetition, unless $k = 0$.
Deduce that the sequence has exactly m different digits in it and
that these m digits are continually repeated in the same order.
These m digits form one of the cycles of α.

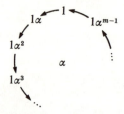

16 The next step is to show that, for any $\alpha \in S_n$, the cycles formed as in
qn 15 are either identical or disjoint. Let a and b be any two of the
digits $1, 2, 3, \ldots , n$, and consider the two sequences

$$a, a\alpha, a\alpha^2, a\alpha^3, \ldots , \text{and}$$

$$b, b\alpha, b\alpha^2, b\alpha^3, \ldots .$$

Suppose that these sequences are not disjoint, and that, in par-
ticular, $a\alpha^i = b\alpha^j$. If $i \geqslant j$ identify the second sequence within the
first and, using qn 15, the first sequence within the second.

17 Use qn 15 and qn 16 to prove that every permutation in S_n is a product
of disjoint cycles.

Signature

We explore the expression of permutations as products of transpositions.

18 Write out the following products in cycle form:

$(12)(13)$, $(12)(13)(14)$, $(12)(13)(14)(15)$, $(12)(13) \ldots (1n)$.

19 Express each of the following permutations as a product of transpositions:

(234), (2468), (13579), $(12)(345)(6789)$, $\begin{pmatrix} 1 & 2 & 3 & 4 & 5 & 6 & 7 & 8 & 9 \\ 6 & 9 & 8 & 3 & 1 & 5 & 7 & 4 & 2 \end{pmatrix}$.

20 How can you be sure that *any* permutation in S_n can be expressed as a product of transpositions? Can the identity be so expressed?

21 Express the 3-cycle (123) in three different ways as a product of two transpositions. Express the same 3-cycle as a product of four transpositions.

The questions that follow lead towards the result that a permutation is *either* the product of an even number of transpositions *or* the product of an odd number of transpositions *but not both*.

22 Evaluate the product $(123456789)(16)$.

23 Evaluate the product $(a_1a_2 \ldots a_rb_1b_2 \ldots b_s)(a_1b_1)$.

24 Evaluate the product $(123)(45678)(14)$.

25 Evaluate the product $(a_1a_2 \ldots a_r)(b_1b_2 \ldots b_s)(a_1b_1)$.

26 If α is a permutation in S_n which is a product of c disjoint cycles (singleton cycles are to be counted) and τ is a transposition in S_n, prove that $\alpha\tau$ is a product of either $c + 1$ or $c - 1$ disjoint cycles.

27 If τ_1 is a transposition in S_n, check that τ_1 consists of $n - 1$ cycles and that $n - 1$ is even or odd according as $n + 1$ is even or odd. If τ_1, $\tau_2, \ldots, \tau_m$ are transpositions in S_n, prove by induction on m that when the product $\tau_1\tau_2 \ldots \tau_m$ is expressed in disjoint cycles, the number of disjoint cycles is even or odd according as $n + m$ is even or odd.

28 If τ_1, $\tau_2, \ldots, \tau_m$ and σ_1, $\sigma_2, \ldots, \sigma_k$ are all transpositions in S_n and $\tau_1\tau_2 \ldots \tau_m = \sigma_1\sigma_2 \ldots \sigma_k$, prove that m and k are both even or both odd.

29 An element of S_n which may be expressed as a product of an even number of transpositions is called an *even permutation*. An element of S_n which may be expressed as a product of an odd number of transpositions is called an *odd permutation*. The *signature* of a permutation is $+1$ for even permutations and -1 for odd permutations. Prove that if $\alpha, \beta \in S_n$ then the signature of $\alpha\beta$ is equal to (signature of α) $\cdot$ (signature of β).

30 What is the signature of (1234) and of (12345)? What is the signature of (123 . . . *n*)?

31 Determine the signature of each of the elements of S_4.

32 Write down the elements of the group generated by (123456) and determine the signature of each element.

33 Write down the elements of the group generated by (12345) and determine the signature of each element.

Alternating group

34 The set of even permutations in S_n is denoted by A_n.
Prove that the product of two elements of A_n is in A_n.

   In the case of finite sets of permutations this is sufficient to ensure that all four of the properties in qn 1.34 hold for A_n, which is called the *alternating group*.

35 List the elements of A_3 and the elements of A_4.

36 Denote the set of odd permutations in S_n by B_n. Show that the function of S_n to itself given by $\alpha \mapsto \alpha \cdot (12)$ is a bijection, and deduce that the number of elements in A_n equals the number of elements in B_n.

37 Does $\begin{pmatrix} 1 & 2 & 3 & 4 & 5 & 6 \\ 4 & 3 & 6 & 1 & 5 & 2 \end{pmatrix}$ = (34)(45)(23)(12)(56)(23)(45)(34)(23)?

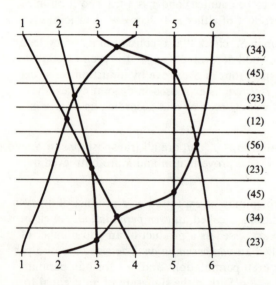

Subgroups of S_n

Any subset of S_A which satisfies the four conditions of qn 1.34 is called a group and is a subgroup of S_A. One way of finding subgroups is to choose some property or structure in A and to consider the subset of S_A which fixes this property.

38 List those elements of S_4 for which $4 \mapsto 4$. This set may be defined as $\{\alpha | \alpha \in S_4, 4\alpha = 4\}$ and is called the *stabiliser* of 4 in S_4. Compare this set with S_3 and describe the similarities.

39 List the elements of S_3 which stabilise 1, 2 and 3 respectively. These three lists form the sets
$\{\alpha | \alpha \in S_3, 1\alpha = 1\}$, $\{\alpha | \alpha \in S_3, 2\alpha = 2\}$ and $\{\alpha | \alpha \in S_3, 3\alpha = 3\}$.
Do each of these sets satisfy the four properties of a group as stated in qn 1.34?

40 Let a be an element of the set A and let
$T = \{\alpha | \alpha \in S_A, a\alpha = a\}$.
Does T satisfy the four properties of a group as stated in qn 1.34? The conclusion we draw is that T is a group, a *subgroup* of S_A, known as the *stabiliser* of a.

41 List and name the set of elements of S_4 under which the set $\{1, 2, 3\}$ is mapped onto itself.

42 Can you match the elements of the set T in qn 40 with the elements of $S_{A-\{a\}}$?

43 List those functions in S_4 which stabilise both 3 and 4, that is to say, the elements of the set
$\{\alpha | \alpha \in S_4, 3\alpha = 3, 4\alpha = 4\}$.
Does this set satisfy the four properties of a group listed in qn 1.34? How does this set relate to the stabiliser of 3 and the stabiliser of 4 in S_4?

44 If T is the stabiliser of a in S_A and R is the stabiliser of b in S_A, does the set $T \cap R$ satisfy the conditions for a group as given in qn 1.34? Suggest a name for the set $T \cap R$.

45 Can you match the elements of the set $T \cap R$ in qn 44 with the elements of $S_{A-\{a,b\}}$?

46 List those elements of S_4 which map the set $\{1, 3\}$ onto itself; that is, those elements which either stabilise both 1 and 3 or else interchange these two digits.

By labelling the vertices of a rhombus 1, 2, 3, and 4, name geometric symmetries of the rhombus which correspond with these listed elements of S_4.

These elements of S_4 are said to *fix* the set $\{1, 3\}$. It is important to note that they do not necessarily stabilise either 1 or 3. If $B = \{1, 3\}$, we may describe these elements as forming the set

$$\{\alpha | \alpha \in S_4, B\alpha = B\}.$$

47 Let B be a subset of the set A and let

$$T = \{\alpha | \alpha \in S_A, B\alpha = B\}.$$

Does T satisfy the four conditions for a group given in qn 1.34? The conclusion we draw is that T is a group, the subgroup of S_A, *fixing B*.

48 If $\alpha = (1234)$, then $\{1, 3\}\alpha = \{1\alpha, 3\alpha\} = \{2, 4\}$, and $\{2, 4\}\alpha = \{2\alpha, 4\alpha\} = \{3, 1\} = \{1, 3\}$. If $\beta = (123)$, then $\{1, 3\}\beta = \{1\beta, 3\beta\} = \{2, 1\}$ and $\{2, 4\}\beta = \{2\beta, 4\beta\} = \{4, 3\}$. Thus α preserves the partition of $\{1, 2, 3, 4\}$ into the subsets $\{1, 3\}$ and $\{2, 4\}$ while β does not. List all the elements of S_4 which preserves this partition. Do these permutations form a group?

49 By labelling the vertices of a square 1, 2, 3 and 4, name a geometric symmetry of a square which corresponds to each of the permutations in S_4 which occurs in the list of qn 48.

50 If x_1, x_2, x_3 and x_4 are real numbers, for which elements α of S_4 must $x_1 x_3 + x_2 x_4 = x_{1\alpha} x_{3\alpha} + x_{2\alpha} x_{4\alpha}$ irrespective of the numbers assigned to x_1, x_2, x_3 and x_4. These permutations are said to preserve the value of the expression $x_1 x_3 + x_2 x_4$. Do they form a group?

51 Which elements $\alpha \in S_3$ have the property that

$$(x_1 - x_2)(x_1 - x_3)(x_2 - x_3) = (x_{1\alpha} - x_{2\alpha})(x_{1\alpha} - x_{3\alpha})(x_{2\alpha} - x_{3\alpha})$$

for distinct real numbers x_1, x_2 and x_3?

Cross-ratio

52 If x_1, x_2, x_3 and x_4 are distinct real numbers, find the four elements $\alpha \in S_4$ such that

$$\frac{x_3 - x_1}{x_3 - x_2} \bigg/ \frac{x_4 - x_1}{x_4 - x_2} = \frac{x_{3\alpha} - x_{1\alpha}}{x_{3\alpha} - x_{2\alpha}} \bigg/ \frac{x_{4\alpha} - x_{1\alpha}}{x_{4\alpha} - x_{2\alpha}}.$$

The function we are studying here is called the *cross-ratio* of the four numbers x_1, x_2, x_3, x_4.

53 By labelling the vertices of a rectangle 1, 2, 3, and 4, name a geometric symmetry of a rectangle which corresponds to each of the permutations you found in qn 52.

Summary

Definition   A bijection $A \to A$ is called a *permutation* of A.
qn 1

Notation   When $A = \{1, 2, \ldots, n\}$, S_A is denoted by S_n.
qn 2

Theorem   Every permutation of a finite set can be expressed as a
qn 17   product of disjoint cycles.

Theorem   Every permutation of a finite set can be expressed as a
qns 20, 29   product of transpositions. Neither the transpositions nor
the number of them is unique. No permutation which
can be expressed as a product of an even number of
transpositions can be expressed as a product of an odd
number of transpositions.

Definition   A permutation which can be expressed as a product of
qn 29   an even number of transpositions is called an *even permutation*. A permutation which can be expressed as a
product of an odd number of transpositions is called an
odd permutation.

Definition   The *signature* of a permutation which may be expressed
qn 29   as a product of n transpositions is $(-1)^n$.

Definition   The set of even permutations in S_n is denoted by A_n and
qn 34   called the *alternating group*.

Definition   If $a \in A$, the subset of S_A of elements which fix a is
qn 40   called the *stabiliser* of a. Any such stabiliser is a subgroup of S_A.

Historical note

In 1770, J. L. Lagrange studied the groups S_2, S_3 and S_4 in relation
to the solutions of equations of degree 2, 3, and 4. In 1799, P. Ruffini
proved by induction that there were $n!$ permutations in S_n. In his work
on determinants, A. L. Cauchy had defined the signature of a permuta-
tion using the function $\Pi_{1 \leqslant i < j \leqslant n}(x_i - x_j)$ in 1812. Cauchy devised the
cycle notation and the term transposition and in his work on permuta-
tions in 1815 distinguished between even and odd permutations in the
way we have done in questions 18–33 and identified the alternating
group. It was E. Galois in 1830 who recognised that stabilisers gave
subgroups.

Answers to chapter 2

1 $\begin{pmatrix} 1 & 2 & 3 \\ 1 & 2 & 3 \end{pmatrix}, \begin{pmatrix} 1 & 2 & 3 \\ 2 & 3 & 1 \end{pmatrix}, \begin{pmatrix} 1 & 2 & 3 \\ 3 & 1 & 2 \end{pmatrix}, \begin{pmatrix} 1 & 2 & 3 \\ 1 & 3 & 2 \end{pmatrix}, \begin{pmatrix} 1 & 2 & 3 \\ 3 & 2 & 1 \end{pmatrix}, \begin{pmatrix} 1 & 2 & 3 \\ 2 & 1 & 3 \end{pmatrix} \cdot \begin{pmatrix} 1 & 2 & 3 & 4 \\ 2 & 1 & 3 & 4 \end{pmatrix},$

$\begin{pmatrix} 1 & 2 & 3 & 4 \\ 2 & 3 & 4 & 1 \end{pmatrix}, \begin{pmatrix} 1 & 2 & 3 & 4 \\ 2 & 4 & 1 & 3 \end{pmatrix}, \begin{pmatrix} 1 & 2 & 3 & 4 \\ 2 & 1 & 4 & 3 \end{pmatrix}, \begin{pmatrix} 1 & 2 & 3 & 4 \\ 2 & 4 & 3 & 1 \end{pmatrix}, \begin{pmatrix} 1 & 2 & 3 & 4 \\ 2 & 3 & 1 & 4 \end{pmatrix}.$

2 $\begin{pmatrix} 1 & 2 & 3 \\ 3 & 2 & 1 \end{pmatrix}$ and $\begin{pmatrix} 1 & 2 & 3 \\ 2 & 1 & 3 \end{pmatrix}$ are reflections. $\begin{pmatrix} 1 & 2 & 3 \\ 3 & 1 & 2 \end{pmatrix}$ is a rotation.

3 $\alpha^2 = \begin{pmatrix} 1 & 2 & 3 & 4 & 5 \\ 3 & 4 & 5 & 1 & 2 \end{pmatrix}, \alpha^3 = \begin{pmatrix} 1 & 2 & 3 & 4 & 5 \\ 4 & 5 & 1 & 2 & 3 \end{pmatrix}, \alpha^4 = \begin{pmatrix} 1 & 2 & 3 & 4 & 5 \\ 5 & 1 & 2 & 3 & 4 \end{pmatrix}, \alpha^5 = I, \alpha^6 = \alpha.$

$i = 0, 1, 2, \ldots \quad 1\alpha^i = 1, 2, 3, 4, 5, 1, 2, \ldots,$

$2\alpha^i = 2, 3, 4, 5, 1, 2, 3, \ldots,$

$3\alpha^i = 3, 4, 5, 1, 2, 3, 4, \ldots.$

4 $\alpha^2 = (13524), \alpha^3 = (14253), \alpha^4 = (15432).$

5 $a = 3, b = 5, c = 4, d = 1, e = 2.$

6 $a = 5, b = 2, c = 4.$

7 $(\alpha\alpha \ldots \alpha)(\alpha\alpha \ldots \alpha) = (\alpha\alpha \ldots \alpha)(\alpha\alpha \ldots \alpha).$
$\quad n \text{ times} \quad m \text{ times} \quad\quad m \text{ times} \quad n \text{ times}$

8 $\alpha = (134)(25) = (341)(52) = (25)(134).$

9 (i) (13524), (ii) (1753)(264), (iii) (16)(2)(35847).

10 (1)(2)(3), (123), (132), (1)(23), (13)(2), (12)(3).

11 (1), (12), (13), (14), (23), (24), (34), (123), (132), (124), (142), (134), (143), (234), (243), (1234), (1432), (1324), (1423), (1243), (1342), (12)(34), (13)(24), (14)(23).

12 $\alpha^2 = (135)(246), \alpha^3 = (14)(25)(36), \alpha^4 = (153)(264), \alpha^5 = (165432).$

13 (123), (132), no.

14 Yes, yes.

15 There are only n digits being permuted. By $1\alpha^n$, $n + 1$ images have occurred. $1, 1\alpha, \ldots, 1\alpha^{m-1}$ are all distinct, but $1\alpha^m = 1$, so $1\alpha^{m+i} = 1\alpha^i$.

16 If $a\alpha^i = b\alpha^j$, then $a\alpha^{i-j} = b$ and $a\alpha^{i-j+k} = b\alpha^k$. If $a\alpha^m = a$, then $b\alpha^{j-i+m} = a$.

17 From qn 15, every digit is in a cycle. From qn 16, the cycles are identical or disjoint.

18 (123), (1234), (12345), (123 ... n).

19 (23)(24), (24)(26)(28), (13)(15)(17)(19), (12)(34)(35)(67)(68)(69).
(165)(29)(384)(7) = (16)(15)(29)(38)(34).

20 From qn 17, every permutation is a product of cycles. As in qn 18, every cycle is a product of transpositions.

21 (12)(13) = (23)(21) = (31)(32) = (12)(13)(13)(13).

22 (12345)(6789).

23 $(a_1 \ldots a_r)(b_1 \ldots b_s)$.

24 (12345678).

25 $(a_1 a_2 \ldots a_r b_1 b_2 \ldots b_s)$.

26 Let $\tau = (ab)$. Then either a and b lie in the same cycle of α and, as in qn 23, the number of cycles in $\alpha\tau$ is $c + 1$ or a and b lie in different cycles of α and, as in qn 25, the number of cycles in $\alpha\tau$ is $c - 1$.

27 Suppose the number of disjoint cycles in $\tau_1\tau_2 \ldots \tau_m$ is even or odd according to $n + m$ is even or odd, then from qn 26 $\tau_1\tau_2 \ldots \tau_m\tau_{m+1}$ has one more or one less cycle and so the number of disjoint cycles is even or odd according as $n + m + 1$ is even or odd.

28 From qn 27, $n + m$ and $n + k$ are both even or both odd.

29 If α is a product of m transpositions, the signature of α is $(-1)^m$.

30 sgn (1234) = sgn (12)(13)(14) = $(-1)^3 = -1$.
sgn (12345) = sgn (12)(13)(14)(15) = $(-1)^4 = +1$.
sgn (12 $\ldots$ n) = sgn (12)(13) $\ldots$ ($1n$) = $(-1)^{n-1}$.

31 $+1$ for identity, 3-cycles and pairs of transpositions. -1 for transpositions and 4-cycles.

32 $+1$ for identity, (135)(246) and (153)(264). -1 for (123456), (14)(25)(36) and (165432).

33 All are even permutations.

34 Follows from qn 29.

35 A_3 = identity and two 3-cycles. A_4 = identity, eight 3-cycles and three pairs of transpositions.

36 $\alpha(12) = \beta(12) \Rightarrow \alpha = \beta$. So the function is a bijection. α even implies $\alpha(12)$ odd and α odd implies $\alpha(12)$ even, so the bijection interchanges the sets A_n and B_n.

37 Yes.

38 The stabiliser of 4 in S_4 is (1)(2)(3)(4), (123)(4), (132)(4), (1)(23)(4), (12)(3)(4),. (13)(2)(4) which is the same as S_3 but for the (4) in each permutation.

39 Yes.

40 $a\alpha = a$ and $a\beta = a$ imply $a\alpha\beta = a$. $aI = a$ always and if $a\alpha = a$, $a\alpha\alpha^{-1} = a\alpha^{-1}$ so $a = a\alpha^{-1}$.

41 The stabiliser of 4.

42 Yes.

43 Argue as in qn 40. Stabiliser of 3 and 4 = intersection of the stabilisers of 3 and of 4.

44 Yes, argue as in qn 40.

45 Yes.

46 (1)(2)(3)(4), (1)(3)(24), (13)(2)(4), (13)(24).
Identity, two reflections and half-turn.

47 Yes. Argue as in qn 40.

48 (1), (13), (24), (13)(24), (1234), (12)(34), (1432), (14)(23).

49 Four rotations, including the identity, and four reflections.

50 Same set as qn 48.

51 Only (1), (123) and (132).

52 (1), (12)(34), (13)(24), (14)(23).

53 Identity, half-turn and two reflections.

3

Groups of permutations of $\mathbb{R}$ and $\mathbb{C}$
Isometries of the plane

In this chapter, we shift our attention from the finite sets of chapter 2 to the geometry of the line and the plane. The set of real numbers, denoted by $\mathbb{R}$, is the formal mathematical system for which the set of points of an infinite straight line is the geometric model. Addition of real numbers corresponds to translation in the geometrical model and multiplication to enlargement (or shrinking) in the geometrical model.

The points of the Euclidean plane are commonly denoted by ordered pairs, (x, y), of real numbers. When such ordered pairs are furnished with addition in the form

$$(x_1, y_1) + (x_2, y_2) = (x_1 + x_2, y_1 + y_2)$$

and multiplication in the form

$$(x_1, y_1) \times (x_2, y_2) = (x_1 x_2 - y_1 y_2, x_1 y_2 + x_2 y_1),$$

the algebra of these ordered pairs is precisely the same as that of numbers in the form $x + iy$ where $i^2 = -1$. Numbers of this form are called *complex numbers* and we write $\mathbb{C} = \{x + iy | x, y \in \mathbb{R}\}$. The set of complex numbers is the formal mathematical system for which the set of points of a plane is the geometric model.

Because the sets $\mathbb{R}$ and $\mathbb{C}$ are infinite, it is impossible to make a complete list of the elements of $S_\mathbb{R}$ or $S_\mathbb{C}$. However, there are some elements of these groups which can be conveniently expressed in algebraic form and this makes it possible to identify the subgroups preserving certain structures on a line or in a plane. We will devote most of this chapter to the subgroup of $S_\mathbb{C}$ which gives the distance-preserving mappings of the plane. A function α is said to be *distance-preserving* when, for any two points a and b, the distance from $a\alpha$ to $b\alpha$ is equal to the distance from a to b.

When both the domain and the codomain of a function are geometrical the function may be referred to as a *transformation*.

Concurrent reading: Ledermann, chapters 1, 2 and 3; Knopp, chapters 1 and 2; Pedoe, chapter 5; Martin, chapters 3–9 and 13; Lockwood and Macmillan, chapters 1 and 14.

The real line, $\mathbb{R}$

When the set A is infinite, the cycle notation we developed in chapter 2 is not generally available for the description of elements of S_A, as we can see by letting $A = \mathbb{R}$ and by trying to write down cycles for the permutation $x \mapsto x + 1$. However we can still find subgroups of $S_\mathbb{R}$ by looking for subsets which leave a particular property or structure unaltered.

1 Illustrate $x \mapsto x + 1$ as a mapping of the real line. If, for some given real number a, $\alpha : x \mapsto x + a$ is an element of $S_\mathbb{R}$, prove that, for any two real numbers x and y, $x - y = x\alpha - y\alpha$. Under the transformation α, the segment of the line from x to y has the same length and the same direction as the segment of the line from $x\alpha$ to $y\alpha$, and such a transformation is called a *translation* of $\mathbb{R}$. If

$$T = \{\alpha | \alpha \in S_\mathbb{R}, x - y = x\alpha - y\alpha\},$$

must T be a subgroup of $S_\mathbb{R}$? That is, must T satisfy the conditions of qn 1.34? If $\alpha \in T$ and $0\alpha = a$, prove that $x\alpha = x + a$, so that T consists of translations and every translation of $\mathbb{R}$ has this form.

2 Illustrate $x \mapsto - x + 2$ as a mapping of the real line. If for some given real number a, $\alpha : x \mapsto -x + a$ is an element of $S_\mathbb{R}$, prove that, for any two real numbers x and y, $- (x - y) = x\alpha - y\alpha$. Under the transformation α, the segment of the line from x to y has the same length as the segment of the line from $x\alpha$ to $y\alpha$, but is in the opposite direction. Such an element of $S_\mathbb{R}$ is called a *half-turn* of $\mathbb{R}$. Find the fixed point of α. For both half-turns and translations of $\mathbb{R}$ the absolute values of lengths are preserved. The set of all transformations which preserve the absolute value of lengths is the set of *isometries* of $\mathbb{R}$. If

$$M = \{\alpha | \alpha \in S_\mathbb{R}, |x - y| = |x\alpha - y\alpha|\},$$

must M be a subgroup of $S_\mathbb{R}$? Does M contain the translation group of $\mathbb{R}$, and all the half-turns of $\mathbb{R}$? We now show that M contains nothing else. If $\alpha \in M$ and $0\alpha = 5$, what can 2α and what can $x\alpha$ be? If $\alpha \in M$ and $0\alpha = a$, prove that $x\alpha = \pm x + a$. If, for given α, $x\alpha = x + a$ and $y\alpha = - y + a$, prove that $|x - y| = |x + y|$ and deduce that x or $y = 0$, so that α is *either* a translation *or* a half-turn.

3 Illustrate $x \mapsto 2x + 1$ as a mapping of the real line. Find its fixed point. Such a mapping is called an enlargement with centre $- 1$

and scale factor 2. If $\alpha : x \mapsto ax + b$ is a mapping of the real numbers and $a \neq 0, 1$, find the fixed point (or centre) of α and describe the transformation geometrically. Compare the ratio $(x - y)/(x - z)$ with the ratio $(x\alpha - y\alpha)/(x\alpha - z\alpha)$ for any three distinct real numbers x, y and z. The set of all transformations which preserve ratios of lengths on the real line is the set of affine transformations or *similarities* of the real line. If

$$A = \left\{ \alpha \mid \alpha \in S_\mathbb{R}, \frac{x - y}{x - z} = \frac{x\alpha - y\alpha}{x\alpha - z\alpha}, \text{ for distinct } x, y, z \in \mathbb{R} \right\}$$

must A be a subgroup of $S_\mathbb{R}$? Does A contain all the translations and half-turns of $\mathbb{R}$? If $\alpha \in A$, $0\alpha = 5$ and $1\alpha = 7$, find $y\alpha$. If $\alpha \in A$, $0\alpha = b$ and $1\alpha = a + b$ with $a \neq 0$, prove that $y\alpha = ay + b$, so that all affine transformations or similarities of the real line take the form $\alpha : x \mapsto ax + b$, where $a \neq 0$.

4 If $x' = ax + b$ and $a \neq 0$, find x in terms of x' and deduce the inverse of the mapping $\alpha : x \mapsto ax + b$.

5 Give the algebraic form of the elements in the stabiliser (see qn 2.40) of 0 in the subgroup of similiarites of $\mathbb{R}$. These are called the enlargements with centre 0.

6 (optional) Let $\mathbb{Q}$ denote the set of rational numbers. The transformations of $\mathbb{Q}$ which preserve the structure of addition are known as the linear transformations of $\mathbb{Q}$ and form the set

$$L = \{ \alpha \mid \alpha \in S_\mathbb{Q}, (x + y)\alpha = x\alpha + y\alpha \}.$$

Must L form a subgroup of $S_\mathbb{Q}$?
If $\alpha \in L$, prove that
(i) $0\alpha = 0$,
(ii) $(-x)\alpha = -(x\alpha)$.
Show that if, moreover, $1\alpha = a$, then $n\alpha = na$ for all integers n, and deduce that for every rational number x, $x\alpha = ax$.

The real plane $\mathbb{R}^2$, the Gauss plane or Argand diagram $\mathbb{C}$

In qns 7–12 we develop just enough of the algebra of complex numbers to serve our geometrical purposes from qn 13 onwards.

7 If we use the mapping $(x, y) \mapsto x + iy$ to label the points of $\mathbb{R}^2$ with the complex numbers $\mathbb{C}$, the x-axis is called the real axis and the y-axis is called the imaginary axis. We define the *modulus* of the complex number $x + iy$ to be the real number $\sqrt{(x^2 + y^2)}$, and denote it by $|x + iy|$. Illustrate the modulus of $x + iy$ in

a diagram, and describe it geometrically. If z and w are complex numbers, prove that $|zw| = |z| \cdot |w|$.

8 If $z = x + iy$, and $z \neq 0$, show in a diagram an angle θ such that $\cos \theta = x/|z|$ and $\sin \theta = y/|z|$. For $0 \leqslant \theta < 2\pi$, we define θ to be the *argument* of z, arg z, so that if $\theta = \arg z$,
$$z = |z| \cdot (\cos \theta + i \sin \theta).$$
Prove that arg zw = arg z + arg w (mod 2π).

9 Find $|\cos \theta + i \sin \theta|$, and determine all complex numbers with modulus 1.

10 Is a complex number uniquely determined by its modulus and argument?

11 Writing $\cos \theta + i \sin \theta = e^{i\theta}$, find $e^{i\theta} \cdot e^{i\phi}$ for real numbers θ and ϕ.

12 If $z = x + iy$, with x and y real, we define $\bar{z} = x - iy$, and call $\bar{z}$ the *conjugate* of z. Prove that $z + \bar{z}$ is a real number (wholly real), that $z - \bar{z}$ is i times a real number (wholly imaginary) and that $|z| = |\bar{z}|$. Illustrate the mapping of $\mathbb{R}^2$ given by $z \mapsto \bar{z}$.
Prove that $\overline{z + w} = \bar{z} + \bar{w}$, and use the mapping $z \mapsto \bar{z}$ to illustrate this. Prove also that $\overline{zw} = \bar{z} \cdot \bar{w}$.

Isometries or symmetries of the plane

The study of isometries is a dynamic version of the study of Euclidean congruence. We focus on the preservation of distance because the preservation of distance implies the preservation of angle. Our method will be to construct algebraic formulae for certain familiar isometries, and then show that combinations of these familiar isometries exhaust the possibilities.

13 Any distance-preserving transformation is called an *isometry*. We begin our study of isometries of the plane with a special case. Illustrate $\alpha : z \mapsto z + 2 + i$ as a transformation of $\mathbb{R}^2$. For any two points of the plane z and w, compare the values of $z - w$ and $z\alpha - w\alpha$. What are the geometric implications of this equality? Use complex numbers and the notation we have established in qn 7 to describe the distance between the points z and w. Is α an isometry?

14 If $\alpha : z \mapsto z + c$ is an element of $S_{\mathbb{C}}$, prove that for any two complex numbers z and w, $z - w = z\alpha - w\alpha$. Such an element of $S_{\mathbb{C}}$ is called a *translation* of $\mathbb{R}^2$. If
$$T = \{\alpha | \alpha \in S_{\mathbb{C}}, z - w = z\alpha - w\alpha\},$$
must T be a subgroup of $S_{\mathbb{C}}$? If $\alpha \in T$ and $0\alpha = c$, prove that

$z\alpha = z + c$ so that T consists entirely of translations of $\mathbb{R}^2$ and is called the *translation group of the plane.*

15 Illustrate $\alpha: z \mapsto e^{i\theta}z$ as a mapping of $\mathbb{R}^2$. This mapping is called a rotation through an angle θ about the centre 0. Verify that $|z - w| = |z\alpha - w\alpha|$, so that rotations about 0 are isometries of the plane.

16 If
$$E = \{\alpha | \alpha \in S_C, |z - w| = |z\alpha - w\alpha|\},$$
must E be a subgroup of S_C? Does E contain the translation group? E is called the *group of isometries* or *Euclidean group* of $\mathbb{R}^2$.

17 Is $\alpha: z \mapsto \bar{z}$ an isometry? Identify the line of fixed points of α, and describe a point and its image under α in relation to the line of fixed points. The transformation $z \mapsto \bar{z}$ is called a *reflection in the real axis.* The reflection $z \mapsto \bar{z}$ and the identity, are both isometries fixing the points 0 and 1. In the next question we establish that these two are the only isometries with this effect.

18 Let α be an isometry which stabilises both 0 and 1. Let z be any point of the plane distinct from both 0 and 1. Sketch the circle with centre 0 passing through z and the circle centre 1 passing through z. Must these circles be mapped onto themselves by α? Where are the possible images of z? Deduce that every point on the real axis is fixed by α. If $z\alpha = \bar{z} \neq z$, and $w\alpha = w$, deduce that $|z - w| = |\bar{z} - w|$, so that w lies on the perpendicular bisector of z and $\bar{z}$, and w is real. Deduce that either $z\alpha = z$ for all z, or $z\alpha = \bar{z}$ for all z.

19 Now we use our exact knowledge of isometries fixing 0 and 1 (qn 18) and our partial knowledge of isometries fixing 0 (qn 15) to determine completely the isometries fixing 0.
Let α be an isometry stabilizing 0. Prove that $|1\alpha| = 1$. Deduce that $1\alpha = e^{i\theta}$ for some θ. If $\beta: z \mapsto e^{i\theta}z$, show that $\alpha\beta^{-1}$ fixes both 0 and 1, and deduce that either $\alpha: z \mapsto e^{i\theta}z$ or $\alpha: z \mapsto e^{i\theta}\bar{z}$.

20 We already know that $z \mapsto e^{i\theta}z$ is a rotation about 0. We examine the isometry $z \mapsto e^{i\theta}\bar{z}$ with a view to describing it geometrically.
For any real number r find the image of the complex number $re^{\frac{1}{2}i\theta}$ under the isometry $\alpha: z \mapsto e^{i\theta}\bar{z}$, and identify a line of fixed points of α. Find the image of the complex number $re^{i\phi}$ under α and describe a point and its image under α in relation to the line of fixed points of α.

When an isometry has a line of fixed points which is the perpendicular bisector of the line joining any other point to its image,

3 Groups of permutations of ℝ and ℂ

it is called a *reflection*. The line of fixed points is called the *axis* of the reflection. What is α^2?

21 Express the rotation $z \mapsto e^{i\theta}z$ as a product of two reflections with axes through the centre of the rotation. What is the relation between the angle of the rotation and the angle between the axes? What is the product of these two reflections taken in the reverse order to that which gave $z \mapsto e^{i\theta}z$?

22 Now we use our exact knowledge of isometries fixing 0 (qn 19) and our partial knowledge of isometries which do not fix 0 (qn 14) to completely determine all isometries.
If α is any isometry of the plane, $0\alpha = c$ and τ is the translation $z \mapsto z + c$, use qn 19 to show that $\alpha\tau^{-1}$ is either $z \mapsto e^{i\theta}z$ or $z \mapsto e^{i\theta}\bar{z}$ and deduce that either $\alpha : z \mapsto e^{i\theta}z + c$ or $\alpha : z \mapsto e^{i\theta}\bar{z} + c$.

23 We examine the types of isometry which have emerged from qn 22 and start by considering the isometry $\alpha : z \mapsto e^{i\theta}z + c$.
Name α when $e^{i\theta} = 1$.
If $e^{i\theta} \neq 1$, find a fixed point p of α, and prove that $z\alpha - p = e^{i\theta}(z - p)$. Illustrate this equation geometrically and describe the transformation α. Use qn 4 to find α^{-1}. If $\alpha : z \mapsto e^{i\theta}z + c$ and $\beta : z \mapsto e^{i\phi}z + d$, calculate $\alpha\beta$.
Describe the composite of two rotations geometrically, and determine when this is a translation.

24 If ABC is a triangle we may denote the reflection in BC by α, the reflection in CA by β and the reflection in AB by γ. Using the ideas of qn 21, describe the transformation $\beta\gamma$ geometrically, and also describe the transformation $\gamma\alpha$. Use the equation $(\beta\gamma)(\gamma\alpha) = \beta\alpha$ to illustrate the last sentence of qn 23.

This completes our study of the transformation $z \mapsto e^{i\theta}z + c$ and of rotations and translations for the time being.

25 If $\alpha : z \mapsto e^{i\theta}\bar{z} + c$, find the condition on $e^{i\theta}$ and c which makes α^2 the identity.

26 Let $\alpha : z \mapsto e^{i\theta}\bar{z} + c$.
  (i) If α has a fixed point prove that $e^{i\theta}\bar{c} + c = 0$.
  (ii) If $e^{i\theta}\bar{c} + c = 0$ prove that the midpoint of 0 and 0α is fixed by α.
  (iii) If $(\frac{1}{2}c)\alpha = \frac{1}{2}c$ prove that $(\frac{1}{2}c + re^{\frac{1}{2}i\theta})\alpha = \frac{1}{2}c + re^{\frac{1}{2}i\theta}$ for all real numbers r.
  (iv) If $e^{i\theta}\bar{c} + c = 0$, show that the axis of the reflection $z \mapsto e^{i\theta}\bar{z}$ is perpendicular to the translation $z \mapsto z + c$.

27 If an isometry α has a line of fixed points, use the ideas of qn 18 to prove that α is either the identity or a reflection with the line of fixed points as axis.

28 Illustrate the isometry $z \mapsto \bar{z} + 1$ by drawing a triangle and its image. Is this transformation a translation, a rotation or a reflection? Is there a line of the plane which is mapped onto itself by this transformation? Is there just one such line? If $\alpha : z \mapsto \bar{z} + 1$, is there a translation τ which coincides with α on this fixed line? What are the fixed points of $\alpha\tau^{-1}$? What kind of an isometry is $\alpha\tau^{-1}$? What can you say about the location of the midpoint of z and $z\alpha$?

29 For what values of the complex number c does the translation $z \mapsto z + c$ commute with the reflection $z \mapsto \bar{z}$? (See qn 6.29.)

30 When an isometry is the product of a reflection and a translation in the direction of the axis of the reflection it is called a *glide-reflection*. Suppose that $\alpha : z \mapsto e^{i\theta}\bar{z} + c$ is not a reflection.
   (i) Prove that α^2 is a translation.
   (ii) If τ is the translation such that $\tau^2 = \alpha^2$ prove that $\alpha\tau = \tau\alpha$.
   (iii) Prove that $\alpha\tau^{-1}$ has order 2 (from (ii)).
   (iv) Prove that $\alpha = \varrho\tau$ for some reflection ϱ (from qns 25, 26 and 27) and that $\varrho\tau = \tau\varrho$.
   (v) Prove that if A is a point which is fixed by ϱ, then $A\tau$ is also fixed by ϱ, so that α is a glide-reflection.
   (vi) If ϱ is a reflection and τ a translation such that $\alpha = \varrho\tau = \tau\varrho$, by considering α^2, prove that τ and ϱ are uniquely defined by α.

We have now completed the proof that every isometry of the plane is either a translation, a rotation, a reflection or a glide-reflection. In qn 31 we analyse these four types of isometry in terms of their fixed points and in qns 32–38 in terms of their effects on angles.

31 (i) What isometries have no fixed points?
   (ii) What isometries have exactly one fixed point?
   (iii) What isometries have more than one fixed point?

32 If z and w are distinct nonzero points, illustrate arg (z/w).

33 If z, w and t are the vertices of a triangle, illustrate $z - w$, $z - t$ and $\arg \dfrac{z - w}{z - t}$.

34 If $\alpha : z \mapsto e^{i\theta}z + c$, show that for any three distinct complex numbers z, w and t
$$\arg \frac{z - w}{z - t} = \arg \frac{z\alpha - w\alpha}{z\alpha - t\alpha}.$$

35 If $z \neq 0$, what is the relationship between arg z and arg $\bar{z}$?

36 If $\alpha : z \mapsto e^{i\theta}\bar{z} + c$, show that for any three distinct complex numbers z, w and t

$$\arg \frac{z - w}{z - t} = 2\pi - \arg \frac{z\alpha - w\alpha}{z\alpha - t\alpha}.$$

37 Using the notation of qn 16, we define

$$D = \left\{ \alpha | \alpha \in E, \arg \frac{z - w}{z - t} = \arg \frac{z\alpha - w\alpha}{z\alpha - t\alpha} \right\}.$$

Must D be a subgroup of E? Identify the elements of D. D is known as the group of *direct* isometries of the plane. What can you claim about the product of a rotation and a translation?

38 If $\alpha : z \mapsto e^{i\theta}\bar{z} + c$, α is known as an *opposite* isometry. Describe the types of opposite isometry of the plane, and prove that the product of any two opposite isometries is direct.

Cyclic and dihedral groups

We have found in qn 19 that the stabiliser of 0 in the Euclidean group consists of rotations of the form $z \mapsto e^{i\theta}z$ and reflections of the form $z \mapsto e^{i\theta}\bar{z}$. In qns 39–47 we identify some important finite subgroups.

39 For what values of θ does $e^{i\theta} = 1$ or -1?

40 Find all the direct isometries which fix the point 0 and map the set $\{1, -1\}$ onto itself. The group thus formed is an example of the *cyclic group C_2*.

41 Find all the isometries which fix the point 0 and map the set $\{1, -1\}$ onto itself. The group thus formed is an example of the *dihedral group D_2*.

42 If $\omega = -\frac{1}{2} + \frac{1}{2}i\sqrt{3}$, find ω^2 and ω^3. For what values of θ does $e^{i\theta} = \omega$, ω^2 or ω^3?

43 Find all the direct isometries which fix the point 0 and map the set $\{1, \omega, \omega^2\}$ onto itself. The group thus formed is an example of the *cyclic group C_3*.

44 Find all the isometries which fix the point 0 and map the set $\{1, \omega, \omega^2\}$ onto itself. The group thus formed is an example of the *dihedral group D_3*. By labelling the point ω with the digit 2 and the point ω^2 with the digit 3 compare D_3 with the group of permutations S_3. See qn 2.2.

45 For what values of θ does $e^{i\theta} = i$, i^2, i^3 or i^4?

46 Find all the direct isometries which fix the point 0 and map the set $\{1, i, -1, -i\}$ onto itself. The group thus formed is an example of the *cyclic group* C_4.

47 Find all the isometries which fix the point 0 and map the set $\{1, i, -1, -i\}$ onto itself. The group thus formed is an example of the *dihedral group* D_4. By labelling the point i with the digit 2, the point -1 with the digit 3, and the point $-i$ with the digit 4, find a subgroup of S_4 which is like D_4. See qn 2.49.

Similarities

An isometry maps a triangle to a congruent triangle, and conversely, if two triangles are congruent there is an isometry which maps one onto the other. If we now relax the distance-preserving condition, we may ask what transformations map a triangle to a similar triangle. We will call such a transformation a *similarity* and note that if under such a transformation, $A \mapsto A'$, $B \mapsto B'$ and $C \mapsto C'$, then

$$\frac{AB}{AC} = \frac{A'B'}{A'C'}.$$

48 By examining the triangle 0, 2, i and its image, describe the effect of the transformation $z \mapsto 2iz - 1$.

49 If $\alpha : z \mapsto az + b$ is an element of S_C, prove that for any three distinct complex numbers z, w and t

$$\frac{z - w}{z - t} = \frac{z\alpha - w\alpha}{z\alpha - t\alpha}.$$

Such an element of S_C is called a *direct similarity*.

50 If

$$S = \left\{\alpha | \alpha \in S_C, \frac{z - w}{z - t} = \frac{z\alpha - w\alpha}{z\alpha - t\alpha}\right\},$$

must S be a subgroup of S_C?
If $\alpha \in S$, with $0\alpha = c$ and $1\alpha = a + c$, say why $a \neq 0$, and prove that $t\alpha = at + c$, so that α is a direct similarity.

51 The transformation $z \mapsto 2iz$ is the product of an enlargement $z \mapsto 2z$ with a scale factor 2, and a rotation $z \mapsto iz$.
When the enlargement and the rotation have the same centre such a transformation is called a *spiral similarity*. If $|a| = 0, 1$, is the transformation $z \mapsto az$ a spiral similarity?

52 Find the fixed point of $z \mapsto 2iz - 1$. Is this transformation a spiral similarity?

53 If $|a| \neq 0$ or 1, find a unique point p which is fixed by the direct similarity $\alpha : z \mapsto az + c$. Prove that $z\alpha - p = a(z - p)$. Illustrate this equation and deduce that α is a spiral similarity.
Name the possible forms of direct similarity and distinguish between them algebraically.

Transformations of the type $z \mapsto a\bar{z} + c$ are called *opposite similarities*, and, with the direct similarities, complete the set of similarities of the plane.

Transitive groups of transformations

In the study of permutation groups, the word *transitivity* is used to describe a degree of freedom to move the points or letters about with the transformations of the group in question.

54 If a and b are complex numbers, can you find a translation which maps a and b? Your success here leads us to say that the translation group is *transitive* on the points of the plane. If τ and σ are translations such that $0\tau = a$ and $0\sigma = b$, can you use τ and σ to define the translation mapping a to b?

55 For distinct complex numbers, a and b, construct a direct similarity α such that $0\alpha = a$ and $1\alpha = b$. For distinct complex numbers, c and d, construct a direct similarity γ such that $0\gamma = c$ and $1\gamma = d$. Use α and γ to construct a direct similarity σ such that $a\sigma = c$ and $b\sigma = d$. Your success here leads us to say that the group of direct similarities is *doubly transitive* on the points of the plane.

56 Prove that the translation group is *not* doubly transitive on the points of the plane.

57 Is the Euclidean group
  (i) transitive
  (ii) doubly transitive
on the points of the plane?

58 Using qn 3, determine whether the affine group on $\mathbb{R}$ is
  (i) transitive
  (ii) doubly transitive
on $\mathbb{R}$.

59 If r is a real number $\neq 0, 1$, find the unique fixed point of the transformation $\delta : z \mapsto rz + c$ and if this fixed point is p show that $z\delta - p = r(z - p)$. Describe this equation geometrically, and name δ. In fact the elements α of $S_{\mathbb{C}}$ such that arg $(z - w) =$ arg $(z\alpha - w\alpha)$ all have the form $z \mapsto rz + c$ for real $r \neq 0$ and

form the *dilatation group* of the plane. Is this dilatation group
(i) transitive, (ii) doubly transitive on the points of the plane?

Summary

Definition The *translation group* on $\mathbb{R}$ consists of the elements of
qn 1 $S_\mathbb{R}$ of the form $x \mapsto x + a$.

Definition The *affine group* on $\mathbb{R}$ consists of the elements of $S_\mathbb{R}$ of
qn 3   the form $x \mapsto ax + b$, $a \neq 0$.

Definition The *translation group* of the Euclidean plane consists of
qn 14   those elements of $S_\mathbb{C}$ of the form $z \mapsto z + c$.

Definition The distance-preserving elements of $S_\mathbb{C}$ form the *group*
qn 16   *of isometries* of the plane.

Theorem   All isometries of the plane either have the form
qns 22, $z \mapsto e^{i\theta}z + c$ and are called *direct* isometries, or else
37, 38   have the form $z \mapsto e^{i\theta}\bar{z} + c$ and are called *opposite*
         isometries.

Theorem   Direct isometries are either translations or rotations.
qn 23

Theorem   Opposite isometries are either reflections or glide-
qns 25–27, reflections.
30

Definition The group of *direct similarities* of the Euclidean plane
qns 49, 50 consists of the elements of $S_\mathbb{C}$ of the form $z \mapsto az + c$,
         $a \neq 0$, with a and c complex.

Definition A subgroup of S_A is said to be *transitive* on A when for
qn 54   any pair of points a, $a' \in A$ there exists an element of
         the subgroup mapping $a \mapsto a'$.

Definition A subgroup of S_A is said to be *doubly transitive* on A
qn 55   when for any two distinct points a, $b \in A$ and any two
         distinct points a', $b' \in A$ there exists an element of the
         subgroup mapping $a \mapsto a'$ and $b \mapsto b'$ simultaneously.

Definition The *dilatation group* of the Euclidean plane consists of
qn 59   the elements of $S_\mathbb{C}$ of the form $z \mapsto rz + c$ where r is a
         real number $\neq 0$.

Historical note

Complex numbers were first developed at the end of the seven-
teenth century. The formula $\cos \theta + i \sin \theta = e^{i\theta}$ in its logarithmic
form was due to Cotes (1710) and the formula $(\cos \theta + i \sin \theta)^n = \cos n\theta + \sin n\theta$ to De Moivre (1730). The way in which the complex
numbers represent the points of the plane was described by C. Wessel
(1797), C. F. Gauss (1799) and J. R. Argand (1806). Gauss proved
the fundamental theorem of algebra, that every polynomial equation

has a complex root, in his doctoral dissertation. C. F. Gauss (1831) and W. R. Hamilton (1837) gave the first formal descriptions of complex numbers as ordered pairs of real numbers, and their first formal definition as an algebraic extension of the real numbers is due to A. L. Cauchy (1847). The use of complex numbers to describe transformations of the plane was intensively developed by G. F. B. Riemann from 1851, though he was not particularly concerned with isometries.

Strangely enough, the analysis of isometries of three-dimensional space into translations, rotations, screws and products of these three types with a reflection, by L. Euler (1776), long precedes the corresponding analysis for two dimensions which appears in the work of M. Chasles (1831).

Answers to chapter 3

1 If $\alpha: x \mapsto x + 1$, the 'cycle' starting with 0 contains all the natural numbers. T is a subgroup. If $\alpha\beta \in T$, $x\alpha\beta - y\alpha\beta = x\alpha - y\alpha = x - y$, so $\alpha\beta \in T$. Also $x\alpha^{-1} - y\alpha^{-1} = x\alpha^{-1}\alpha - y\alpha^{-1}\alpha = x - y$, so $\alpha^{-1} \in T$.

2 $\frac{1}{2}a$ is the fixed point of $x \mapsto -x + a$. M is a subgroup, argue as in qn 1.

3 $b/(1 - a)$ is the fixed point of $x \mapsto ax + b$. A is a subgroup, argue as in qn 1. If $0\alpha = 5$ and $1\alpha = 7$, put $x = 0$ and $z = 1$ to get $y\alpha = 2y + 5$. If $0\alpha = b$ and $1\alpha = a + b$, put $x = 0$ and $z = 1$.

4 $x = (x' - b)/a$ so $\alpha^{-1}: x \mapsto x/a - b/a$.

5 $x \mapsto ax$.

6 L is a subgroup, argue as in qn 1. Prove $n\alpha = na$ by induction. If p and q are integers, let $(p/q)\alpha = r'$, then $(q \cdot p/q)\alpha = q \cdot r' = p\alpha = pa$, so $r' = (p/q)a$.

9 If $z = x + iy$ and $|z| = 1$, then $x^2 + y^2 = 1$ and $z = \cos\theta + i\sin\theta$ for some θ.

10 Yes, as in qn 8.

11 From the working of qn 8, $e^{i\theta} \cdot e^{i\phi} = e^{i(\theta + \phi)}$.

12 $z + \bar{z} = 2x$. $z - \bar{z} = 2iy$. $|\bar{z}| = \sqrt{(x^2 + y^2)} = |z|$. $z \mapsto \bar{z}$ is a reflection in the real axis and the image of the parallelogram $0, z, w, z + w$ is again a parallelogram, as $\bar{z} + \bar{w} = \overline{z + w}$. To prove $\overline{zw} = \bar{z} \cdot \bar{w}$, use $z\bar{z} = |z|^2$.

13 $z\alpha - w\alpha = z - w$.

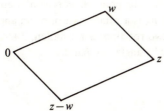

If two complex numbers are equal, then they represent the same point. Now $0, w, z, z - w$ form a parallelogram, so $|z - w| = $ the distance between the points z and w. Thus α is an isometry since $|z - w| = |z\alpha - w\alpha|$.

14 $z\alpha - w\alpha = (z + c) - (w + c) = z - w$. T is a subgroup. Argue as in qn 1.

15 $|z\alpha - w\alpha| = |e^{i\theta}z - e^{i\theta}w| = |e^{i\theta}(z - w)| = |e^{i\theta}| \cdot |z - w| = |z - w|$.

16 E is a subgroup, argue as in qn 1. $E \supset T$.

17 $x + iy = x - iy \Leftrightarrow y = 0$, so the points of the real axis are fixed. If $z\alpha \neq z$ then the real axis is the perpendicular bisector of $zz\alpha$.

18 If an isometry α fixes 0, each circle centre 0 is mapped onto itself. If z is a point of intersection of these two circles, $z\alpha$ is also a point of intersection of these circles. The line of centres (the real axis) is the perpendicular bisector of the common chord. So $z\alpha = z$ or $\bar{z}$. If z lies on the real axis, the two circles touch at z, and z is fixed.

19 If $0\alpha = 0$ then $|1\alpha - 0\alpha| = 1 - 0$ so $|1\alpha| = 1$. From qn 9, $1\alpha = e^{i\theta}$ for some θ. From qn 18, $\alpha\beta^{-1} : z \mapsto z$ or $\alpha\beta^{-1} : z \mapsto \bar{z}$.

20 $re^{i\phi}\alpha = re^{i(\theta - \phi)}$. Now $\tfrac{1}{2}[\phi + (\theta - \phi)] = \tfrac{1}{2}\theta$, so a point and its image are bisected by the line of fixed points.

21 If $\alpha : z \mapsto \bar{z}$ and $\beta : z \mapsto e^{i\theta}\bar{z}$, then $\alpha\beta : z \mapsto e^{i\theta}z$ and $\beta\alpha : z \mapsto e^{-i\theta}z$. Angle between axes of α and β is $\tfrac{1}{2}\theta$.

22 $0\alpha\tau^{-1} = c\tau^{-1} = 0$.

23 When $e^{i\theta} = 1$, $z \mapsto e^{i\theta}z + c$ is a translation (see qn 14). When $e^{i\theta} \neq 1$, $z = e^{i\theta}z + c \Leftrightarrow z = c/(1 - e^{i\theta}) = p$. The isometry α is a rotation through an angle θ with the point p as centre. $\alpha^{-1} : z \mapsto ze^{-i\theta} - ce^{-i\theta}$.
$\alpha\beta : z \mapsto e^{i(\theta + \phi)}z + ce^{i\phi} + d$ which is a rotation through an angle $\theta + \phi$ unless $\theta + \phi = 0$, in which case $\alpha\beta$ is a translation.

24 $\beta\gamma$ is a rotation through an angle $2A$ about the point A. $\gamma\alpha$ is a rotation through an angle $2B$ about the point B. $\beta\alpha$ is a rotation through an angle $2C$ about the point C. The product of two rotations is a rotation, $2A + 2B = 2\pi - 2C$, unless two sides of this 'triangle' are parallel.

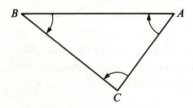

25 $\alpha^2 = 1 \Leftrightarrow e^{i\theta}\bar{c} + c = 0$.

26 (i) If w is a fixed point of α,

$w = e^{i\theta}\bar{w} + c$

so

$\bar{w} = e^{-i\theta}(e^{i\theta}\bar{w} + c) + \bar{c} = \bar{w} + e^{-i\theta}c + \bar{c}.$

and

$e^{-i\theta}c + \bar{c} = 0.$

(ii) $0\alpha = c$, so the midpoint of 0 and 0α is $\frac{1}{2}c$. (iii) If $(\frac{1}{2}c)\alpha = \frac{1}{2}c$ then from (i), $e^{i\theta}\bar{c} + c = 0$. So α has order 2 if and only if α has a line of fixed points. (iv) c and $\bar{c}e^{i\theta}$ are transposed by the reflection and $c = -\bar{c}e^{i\theta}$.

27 Draw circles, centred on the line of fixed points.

28 Not a translation, a rotation or a reflection. The real axis alone is mapped to itself. $\tau : z \mapsto z + 1$. $\alpha\tau^{-1}$ is the reflection in the real axis. If $z\alpha = \bar{z} + 1$, $\frac{1}{2}(z + z\alpha) = \frac{1}{2}(z + \bar{z} + 1)$, which is real.

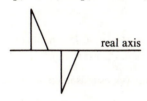

real axis

29 The translation $z \mapsto z + c$ commutes with $z \mapsto \bar{z}$ when $c = \bar{c}$. The complex number c is then purely real and the translation is parallel to the axis of the reflection.

30 (i) If α is not a reflection $e^{i\theta}\bar{c} + c \neq 0$, from qn 25. So $\alpha^2 : z \mapsto z + e^{i\theta}\bar{c} + c$ is a translation. (ii) $\tau : z \mapsto z + \frac{1}{2}(e^{i\theta}\bar{c} + c)$. (iii) $\alpha\tau = \tau\alpha \Rightarrow \tau^{-1}\alpha = \alpha\tau^{-1}$
$\Rightarrow (\alpha\tau^{-1})^2 = \alpha\tau^{-1}\alpha\tau^{-1} = \alpha^2\tau^{-2} = 1.$

(iv) Substitute $\alpha = \varrho\tau$ in $\alpha\tau = \tau\alpha$. (v) $A\varrho = A \Rightarrow A\varrho\tau = A\tau \Rightarrow A\tau\varrho = A\tau$. (vi) $\alpha^2 = (\varrho\tau)^2 = \varrho\tau\varrho\tau = \varrho\varrho\tau\tau = \tau^2$, so τ is uniquely defined, and $\varrho = \alpha\tau^{-1}$.

31 (i) translations and glide-reflections. (ii) rotations. (iii) reflections and the identity.

32

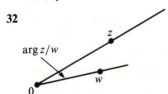

arg z/w

0

33

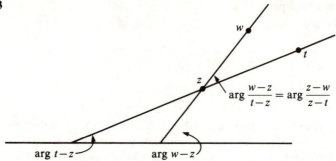

34 $\dfrac{z\alpha - w\alpha}{z\alpha - t\alpha} = \dfrac{z - w}{z - t}$.

35 Arg $z = 2\pi - \arg \bar{z}$.

36 $\dfrac{z\alpha - w\alpha}{z\alpha - t\alpha} = \dfrac{\bar{z} - \bar{w}}{\bar{z} - \bar{t}}$.

37 Arguing as in qn 1, D is a subgroup. From qns 34 and 36 D consists of rotations and translations. It is a rotation through the same angle.

38 The reflections and glide-reflections are the only opposite isometries of the plane.

39 $e^{i\theta} = 1$ when $\theta = 0$. $e^{i\theta} = -1$ when $\theta = \pi$.

40 $z \mapsto z$ and $z \mapsto -z$.

41 $z \mapsto z$, $z \mapsto -z$, $z \mapsto \bar{z}$ and $z \mapsto -\bar{z}$.

42 $\omega^2 = -\frac{1}{2} - \frac{1}{2}i\sqrt{3}$, $\omega^3 = 1$. $\theta = 2\pi/3, 4\pi/3, 0$.

43 $z \mapsto z$, $z \mapsto \omega z$, $z \mapsto \omega^2 z$.

44 $z \mapsto z$, $z \mapsto \omega z$, $z \mapsto \omega^2 z$, $z \mapsto \bar{z}$, $z \mapsto \omega \bar{z}$, $z \mapsto \omega^2 \bar{z}$.

45 $\theta = \frac{1}{2}\pi, \pi, \frac{3}{2}\pi, 0$.

46 $z \mapsto z$, iz, $-z$ and $-iz$.

47 $z \mapsto \pm z, \pm iz, \pm \bar{z}, \pm i\bar{z}$.

48 It has been rotated through $\frac{1}{2}\pi$ and enlarged by a factor 2.

50 S is a subgroup, argue as in qn 3.

51 $a = re^{i\theta}$ so $z \mapsto az$ is the enlargement $z \mapsto rz$ followed by the rotation $z \mapsto e^{i\theta}z$.

52 $1/(2i - 1) = -\frac{1}{5} - \frac{2}{5}i$. If α is a quarter-turn about this point
$z\alpha = iz - \frac{3}{5} - \frac{1}{5}i$.

If β is an enlargement by a factor of 2 from this centre

$z\beta = 2z + \frac{1}{5} + \frac{2}{3}i,$

and $z\alpha\beta = 2iz - 1$, so $z \mapsto 2iz - 1$ is a spiral similarity.

53 $p = ap + c \Leftrightarrow p = c/(1 - a)$. α is a direct similarity being the product of a rotation and an enlargement with centres at p. $a = 1$ gives a translation. $a = e^{i\theta}$ gives a rotation. $|a| \neq 1$ gives a spiral similarity.

54 $z \mapsto z + b - a$. $\tau^{-1}\sigma$.

55 $\alpha : z \mapsto (b - a)z + a; \gamma : z \mapsto (d - c)z + c; \sigma = \alpha^{-1}\gamma$.

56 A translation is uniquely defined by the image of the origin.

57 (i) Transitive because it contains the translation group. (ii) Cannot map $0 \mapsto 0$ and $1 \mapsto 2$ by the same isometry.

58 (i) yes, (ii) yes.

59 $p = c/(1 - r)$. δ is an enlargement by a scale factor r, centre p. (i) Transitive because it contains the translation group. (ii) Cannot map $0 \mapsto 0$ and $1 \mapsto i$ by the same dilatation.

4

The Möbius group

To extend our investigation of transformations of the plane from the similarity group of chapter 3 (with transformations of the form $z \mapsto az + b$) to the Möbius group with transformations of the form $z \mapsto \dfrac{az + b}{cz + d}$, we need to count $z \mapsto 1/z$ as a transformation of the plane, and we can do this if we adjoin a single point '∞' to the plane. This 'completed plane' is then like the surface of a sphere in a sense which we will establish with stereographic projection.

As a preparation for the idea of a 'completed plane' we will start this chapter by considering a real line with a single 'point at infinity' adjoined. This 'completed line' then has a certain similarity to the circumference or a circle, and we can relate some algebraic transformations of the completed line to geometric transformations of the circle.

The symbol '∞' will take on those properties which we give it, and no others are to be presumed. It does not denote a real or complex number, however large.

Concurrent reading: Knopp, chapters 3, 4, and 5; Pedoe, chapter 6; Ford, chapters 1 and 2.

The completed line, $\mathbb{R} \cup \{\infty\}$

1 Let ON be a line segment of unit length and let Σ be the circle on ON as diameter. We define a function π of $\Sigma - \{N\}$ to the tangent at O by $\pi: A \mapsto A'$, where NAA' are collinear. Is π an injection, a surjection, a bijection? The function π is called *projection* from N.

2 If A and B are points of Σ such that AB is parallel to ON but distinct from it, show that angle ONA + angle ONB = one right angle. Use the Euclidean theorem about the angle in a semicircle.

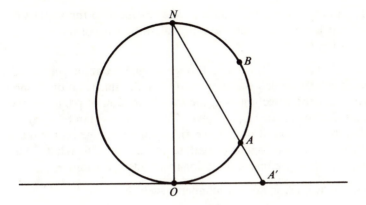

3 If angle $ONA = \theta$, what is the distance OA'?

4 What is $OA' \cdot OB'$, where $B' = B\pi$?

5 If the points on the tangent at O are labelled with their directed distances from O, and the points of $\Sigma - \{N\}$ are given the labels of their images under π, what is the label for B if the label for A is x (where x is not equal to 0)? Use these labels to describe a reflection of Σ about the diameter perpendicular to ON. Ignore the points O and N at first, and then describe what happens to them separately. We now label N with the symbol ∞ and use the algebraic description $x \mapsto 1/x$ to describe the transformation of the whole circumference.

6 Under an anticlockwise rotation through an angle α about the centre of Σ, show that the image of the point of Σ which was labelled $\tan \theta$ in qn 5 is the point which would have been labelled $\tan (\theta + \tfrac{1}{2}\alpha)$. Use the Euclidean theorem that the angle at the centre of the circle is twice the angle at the circumference standing on the same arc. Deduce that for all real points of the circle, except for the point $x = \tan (\tfrac{1}{2}\pi - \tfrac{1}{2}\alpha) = 1/\tan \tfrac{1}{2}\alpha$, this rotation may be expressed in the form

$$x \mapsto \frac{x + a}{1 - xa},$$

where $a = \tan \tfrac{1}{2}\alpha$. What is the image of ∞ under this rotation? What point has ∞ as its image under this rotation? Can you identify these points from the algebraic expression

$$\frac{x + a}{1 - xa}?$$

(It is intuitively useful to imagine what happens to this expression as x gets large, and to decide what values of x make this expression large.)

7 If the diameter ON is rotated through an angle α about the centre of the circle and we call the image of ON under this rotation d, show that under the reflection in d, the transformation of points on the circumference of the circle is given by $\tan \theta \mapsto - \tan (\theta - \alpha)$ using the labelling of qn 5. Give an algebraic expression for this transformation analogous to that at the end of qn 6. What is the image and the pre-image of ∞ for this transformation?

8 Verify that there is no real number x such that

$$\frac{ax + b}{cx + d} = \frac{a}{c},$$

assuming that $c \neq 0$, and $ad - bc \neq 0$. Prove that if the real number r is not equal to a/c, then there is a unique real number x such that

$$\frac{ax + b}{cx + d} = r.$$

What real number x does not have a real image under the transformation

$$x \mapsto \frac{ax + b}{cx + d}?$$

State the image and the pre-image of ∞ if the function

$$x \mapsto \frac{ax + b}{cx + d}$$

is in $S_{\mathbb{R} \cup \{\infty\}}$ and $c \neq 0$.

9 If $c = 0$ and $ad - bc \neq 0$, verify that

$$x \mapsto \frac{ax + b}{cx + d}$$

is an affine transformation of $\mathbb{R}$ (see qn 3.3). If this is to be a function in $S_{\mathbb{R} \cup \{\infty\}}$, what must be the image and pre-image of ∞?

10 Verify that

$$\frac{ax + b}{cx + d} = \frac{ay + b}{cy + d} \Leftrightarrow (ad - bc)(x - y) = 0.$$

so that $ad - bc \neq 0$ is a necessary condition that

$$x \mapsto \frac{ax + b}{cx + d}$$

be an injection. If $c \neq 0$, evaluate the product of the functions

$x \mapsto cx + d,$

$x \mapsto 1/x,$

$x \mapsto \dfrac{bc - ad}{c}x,$

and

$x \mapsto x + a/c$

and deduce that

$x \mapsto \dfrac{ax + b}{cx + d}$

is in $S_{\mathbb{R} \cup \{\infty\}}$ provided that $ad - bc \neq 0$.

Cross-ratio

We now establish that the transformations of qn 10 are precisely those preserving a specific structure on the projective line, and hence form a group.

11 If

$\alpha : x \mapsto \dfrac{ax + b}{cx + d}$ and $ad - bc \neq 0,$

verify that for any distinct real numbers x, y, z and t

$$\dfrac{z - x}{z - y} \bigg/ \dfrac{t - x}{t - y} = \dfrac{z\alpha - x\alpha}{z\alpha - y\alpha} \bigg/ \dfrac{t\alpha - x\alpha}{t\alpha - y\alpha}$$

assuming that none of $x\alpha$, $y\alpha$, $z\alpha$ or $t\alpha = \infty$.

$$\dfrac{z - x}{z - y} \bigg/ \dfrac{t - x}{t - y}$$

is defined to be the *cross-ratio* $R(x, y; z, t)$, and if we further define

$$R(\infty, y; z, t) = \dfrac{t - y}{z - y}, \quad R(x, \infty; z, t) = \dfrac{z - x}{t - x},$$

$$R(x, y; \infty, t) = \dfrac{t - y}{t - x}, \quad R(x, y; z, \infty) = \dfrac{z - x}{z - y},$$

we consider the set

$T = \{\alpha | \alpha \in S_{\mathbb{R} \cup \{\infty\}}, \quad R(x, y; z, t) = R(x\alpha, y\alpha; z\alpha, t\alpha)\}.$

Must T be a subgroup of $S_{\mathbb{R} \cup \{\infty\}}$?

If $\alpha \in T$ and $0\alpha = a_1$, $1\alpha = a_2$ and $\infty\alpha = a_3$, then

$R(x, 0; 1, \infty) = R(x\alpha, a_1; a_2, a_3).$

Deduce that

$$x\alpha = \dfrac{a_3(a_2 - a_1)x + a_1(a_3 - a_2)}{(a_2 - a_1)x + (a_3 - a_2)}.$$

The group of cross-ratio-preserving transformations of the completed line is known as the *projective group* on the line. See chapter 18. The completed line is also known as the projective line.

Triple transitivity

12 If a_1, a_2, a_3 and b_1, b_2, b_3 are triples of distinct points on the real projective line, there is a cross-ratio-preserving transformation α such that $0\alpha = a_1$, $1\alpha = a_2$ and $\infty\alpha = a_3$, and a cross-ratio-preserving transformation β such that $0\beta = b_1$, $1\beta = b_2$ and $\infty\beta = b_3$. Find a cross-ratio-preserving transformation γ such that $a_1\gamma = b_1$, $a_2\gamma = b_2$, and $a_3\gamma = b_3$.

This makes the projective group on the real line *triply transitive*.

Möbius transformation. Möbius plane

13 All the work in this chapter so far has been with real numbers. If a, b, c and d are complex numbers, and

$$z \mapsto \frac{az + b}{cz + d}$$

is a bijection of $\mathbb{C} \cup \{\infty\}$ to itself, then this bijection is called a *Möbius transformation*. The set $\mathbb{C} \cup \{\infty\}$ is sometimes called the complex projective line, but we will generally call it the *Möbius plane*. If

$$z \mapsto \frac{az + b}{cz + d}$$

is a Möbius transformation, find the image and pre-image of ∞ (i) when $c = 0$, (ii) when $c \neq 0$.

14 Assuming that $c \neq 0$, find the product of the four transformations of the Möbius plane

$$z \mapsto cz + d,$$

$$z \mapsto 1/z,$$

$$z \mapsto \frac{bc - ad}{c}z,$$

$$z \mapsto z + a/c.$$

Which of these four are necessarily Möbius transformations? What is the necessary and sufficient condition that the product is a Möbius transformation? When the product is a Möbius transformation, describe the first, third and fourth of these as transformations of the Gauss plane or Argand diagram.

15 If the cross-ratio of four elements of the Möbius plane is defined in a manner strictly analogous to that of qn 11, must every Möbius transformation preserve cross-ratio?

16 By an argument analogous to that of qn 11 show that every cross-ratio-preserving transformation of $\mathbb{C} \cup \{\infty\}$ is a Möbius transformation.

The group of cross-ratio-preserving transformations of the Möbius plane is called the *Möbius group*.

17 Is a Möbius transformation uniquely determined by the images of 0, 1 and ∞?

Is the Möbius group triply transitive on the Möbius plane?

Having seen in qns 13–17 how the real algebra of qns 8–12 can be extended to complex algebra, we identify the corresponding geometrical extension from one to two dimensions.

18 If the projection of qn 1 is extended by setting the figure in three dimensions and rotating through $360°$ about ON, the real line, the tangent at O, is extended to a real plane. To what is $\Sigma - \{N\}$ extended? If the real tangent plane is labelled with $\mathbb{C}$, what model of $\mathbb{C} \cup \{\infty\}$ does the projection imply?

The fact that Möbius transformations preserve complex cross-ratios has striking geometrical consequences for the images of straight lines and circles.

19 For any real number y, show that

$$\left| \frac{1}{2} - \frac{1}{1 + iy} \right| = \frac{1}{2}.$$

What is the image of the straight line $\{z | z = 1 + iy, y \in \mathbb{R}\}$ under the Möbius transformation $z \mapsto 1/z$? What is the image of the circle $\{z \| z - \frac{1}{2}| = \frac{1}{2}\}$ under this Möbius transformation?

20 Using qn 3.33, describe

$$\arg \frac{z - w}{z - t}$$

for complex numbers z, w and t corresponding to the vertices of a triangle in the Gauss plane in terms of the angles zw and zt make with the real axis. If

$$\arg \frac{z - w}{z - t} = \arg \frac{c - w}{c - t},$$

what can be said about the four points z, w, t and c?

Cross-ratios can be used to identify concyclic and collinear points.

21 If z, w, t, and c are concyclic points, what is the value of

$$\arg \frac{z - w}{z - t} \bigg/ \frac{c - w}{c - t}$$

and what does this imply about the complex number

$$\frac{z - w}{z - t} \bigg/ \frac{c - w}{c - t}?$$

22 If z, w, t, and c are collinear points, what is the value of

$$\arg \frac{z - w}{z - t} \bigg/ \frac{c - w}{c - t}?$$

23 If z, w, t, and c are collinear points what is the value of $\arg R(\infty, t; z, c)$, $\arg R(w, \infty; z, c)$, $\arg R(w, t; \infty, c)$ and $\arg R(w, t; z, \infty)$?

24 If z, t, w are distinct complex numbers, describe the subset of the Möbius plane $\{c | \arg R(w, t; z, c) = 0 \text{ or } \pi\}$
  (i) when w, t, $z \in \mathbb{C}$ and represent the vertices of a triangle in the Gauss plane,
  (ii) when w, t, $z \in \mathbb{C}$ and represent collinear points in the Gauss plane.
Also describe this set when one of w, t and z is ∞.

25 What are the possible images of a circle in the Gauss plane under a Möbius transformation?

26 What are the possible images of a straight line with ∞ under a Möbius transformation?

27 If two straight lines in the plane are given, is it possible to find a Möbius transformation which maps one onto the other?

28 If a circle and a straight line in the plane are given, is it possible to find a Möbius transformation which maps the circle onto the line with ∞? (Use triple transitivity.)

The transformation $z \mapsto 1/z$

29 Illustrate the image of $e^{i\theta}$ under $z \mapsto 1/z$. What is the image of $z = 2$ under $z \mapsto 1/z$? What is the image of the circle $|z| = 2$ under this transformation? What is the image of the real line under $z \mapsto 1/z$? What is the image of the line through the origin $\{re^{i\theta} | r \in \mathbb{R}\}$?

30 If r is a real number different from 0, 1, and -1, and $0 < \theta < \pi$, what is the image of the set $\{r, 1/r, e^{i\theta}\}$ under the transformation $z \mapsto 1/z$? Let C denote the circle through the points r, $1/r$ and $e^{i\theta}$. Assuming the secant theorem that if PQ and RS are chords of a circle which intersect at A, then $AP \cdot AQ = AR \cdot AS$, prove that the line joining 0 to $e^{i\theta}$ is a tangent to C.

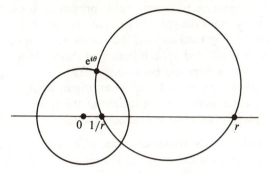

When a radius of one circle is a tangent to another at their point of intersection the two circles are said to cut *orthogonally*. Thus $|z| = 1$ and C cut orthogonally. If C' is the image of C under $z \mapsto 1/z$, what properties of C' can you claim?

31 If a circle Σ with radius a and centre O cuts a circle C orthogonally, and AB is a chord of C which passes through O, prove that $OA \cdot OB = a^2$, by considering OT where T is a point of intersection of Σ and C.

32 If OAB are three points in order on a line and $OA \cdot OB = a^2$, prove that the circle with centre O and radius a is orthogonal to any circle through the two points A and B.

33 If Σ is a circle orthogonal to $|z| = 1$ and Σ meets the real axis at the points r and s, what is rs? What are the images of r and s under the Möbius transformation $z \mapsto 1/z$? What is the image of the circle Σ under the transformation $z \mapsto 1/z$?

The inversion $z \mapsto 1/\bar{z}$

In order to establish the angle-preserving properties of $z \mapsto 1/z$, and hence of all Möbius transformations, we decompose $z \mapsto 1/z$ into the product of the reflection $z \mapsto \bar{z}$ and the, as yet uninvestigated, transformation $z \mapsto 1/\bar{z}$.

34 If $re^{i\theta}$ and $se^{i\phi}$ lie on a line through the origin, what is the relationship between θ and ϕ? If in addition a circle through $re^{i\theta}$ and $se^{i\phi}$ is orthogonal to the circle $|z| = 1$, what is the relationship between the real numbers r and s? Is $re^{i\theta} \cdot \overline{se^{i\phi}}$ independent of r, s, θ and ϕ?

35 If the diagram of qn 1 is rotated in three dimensions about the axis ON, so that the circle on ON as diameter generates a sphere, and the tangent at O generates a tangent plane, the projection from N of the sphere (without N) to the tangent plane is called *stereographic projection*. If the inverse of stereographic projection is used to map the tangent plane onto the sphere (without N), and the complex numbers are thereby used to label the points of the sphere (except for N), while N is labelled ∞, a bijection of the Möbius plane to the surface of the sphere has been constructed.

Taking the diameter of the sphere as 1, and O as origin, what mapping of $\mathbb{C} \cup \{\infty\}$ corresponds to a reflection of the sphere about an equatorial plane parallel to the tangent plane?

36 By considering the product of the transformations $z \mapsto \bar{z}$ and $z \mapsto 1/z$ of the Möbius plane, determine the image of the set of circles and lines of the Möbius plane under the transformation $z \mapsto 1/\bar{z}$. This transformation is known as the *inversion* in the circle $|z| = 1$.

37 If Σ is a circle orthogonal to $|z| = 1$, what is the image of Σ under the inversion $z \mapsto 1/\bar{z}$?

38 Let P be a point on a circle C which does not pass through the origin. Let P' and C' be the images of P and C respectively under the inversion $z \mapsto 1/\bar{z}$. What may be claimed about any circle through P and P' in relation to the circle $|z| = 1$? Let S denote the circle through P and P' which touches C at P. Establish by contradiction that since S and C meet in one point, S and C' may meet in only one point. Thus S touches C' at P'.

39 Let C_1 and C_2 be two circles intersecting at $P \neq 0$. Let C_1', C_2' and P' be the images of C_1, C_2 and P, respectively, under the inversion $z \mapsto 1/\bar{z}$. Let S_1 be the circle through P and P' touching C_1, and let S_2 be the circle through P and P' touching C_2. Why is the angle of intersection of C_1 and C_2 equal to the angle of intersection of S_1 and S_2 and why is this, in turn, equal to the angle of intersection of C_1' and C_2'? It is the absolute values we equate here; the oriented values are in fact reversed. Assume throughout that $P \neq P'$.

40 Investigate the preservation of angles of intersection of circles under the inversion $z \mapsto 1/\bar{z}$
  (i) when one or both of the 'circles' is a straight line,
  (ii) when the circles intersect at 0.

41 If two circles have one point of intersection on $|z| = 1$, why must their angle of intersection be equal to the angle of intersection of their images under the inversion $z \mapsto 1/\bar{z}$?

42 By decomposing an arbitrary Möbius transformation as in qns 14 and 36, show that the angles of intersection of circles and lines are preserved under any Möbius transformation.

This completes our study of the geometric properties which are preserved by Möbius transformations.

Stabilisers

43 Exhibit, in algebraic form

    (i) the stabiliser of ∞,   (v) the stabiliser of 1 and ∞,

    (ii) the stabiliser of 0,   (vi) the stabiliser of 0 and 1

    (iii) the stabiliser of 1,

    (iv) the stabiliser of 0 and ∞,   in the Möbius group.

44 Prove that a Möbius transformation, distinct from the identity, always has either 1 or 2 fixed points.

45 If a Möbius transformation fixes three distinct points, must it be the identity? Use qn 17 to show that there is exactly one Möbius transformation mapping three distinct points to three given images. If $z \mapsto z'$ is a Möbius transformation and

$$\frac{z - a}{z - b} \cdot \frac{c - b}{c - a} = \frac{z' - a'}{z' - b'} \cdot \frac{c' - b'}{c' - a'},$$

what are the images of a, b and c?

46 Find an algebraic condition on a, b, c and d which is necessary and sufficient for

$$z \mapsto \frac{az + b}{cz + d}$$

to have a unique fixed point.

47 Verify that the Möbius transformations which fix only the point ∞, together with the identity, form a group.

Subgroup fixing the upper half plane

48 Show that every Möbius transformation fixing the real line may be expressed in the form

$$z \mapsto \frac{az + b}{cz + d},$$

with a, b, c and d real numbers.

49 If x and y are real numbers and $z = x + \mathrm{i}y$, we write $\mathrm{Im}(z) = y$. If a, b, c and d are real numbers, prove that

$$\mathrm{Im}\left(\frac{az + b}{cz + d}\right) = \frac{\mathrm{Im}\,(z) \cdot (ad - bc)}{|cz + d|^2}.$$

Show that every Möbius transformation fixing the upper half plane

$\{z| \operatorname{Im}(z) > 0\}$ may be expressed in the form

$$z \mapsto \frac{az + b}{cz + d},$$

with a, b, c and d real numbers, and $ad - bc > 0$. Assume that the fixing of the half-plane implies the fixing of its boundary.

Subgroup fixing the unit circle

50 Use qn 32 to show that every circle through the two points w and $1/\bar{w}$ cuts $|z| = 1$ orthogonally. If α is a Möbius transformation which maps the circle $|z| = 1$ onto itself, what can you say about a circle passing through $w\alpha$ and $(1/\bar{w})\alpha$? Deduce that the inversion $z \mapsto 1/\bar{z}$ interchanges $w\alpha$ and $(1/\bar{w})\alpha$.

51 If a Möbius transformation α maps the circle $|z| = 1$ onto itself and fixes the point 0, explain why it must also fix ∞ and so have the form $z \mapsto e^{i\theta}z$ for some θ.

52 If the Möbius transformation

$$z \mapsto \frac{az + b}{cz + d}$$

maps the unit circle $|z| = 1$ to itself, prove from qn 50 that

$$\frac{bz + a}{dz + c} = \frac{\bar{c}z + \bar{d}}{\bar{a}z + \bar{b}}$$

for all complex numbers z. By putting $z = 0$, $- a/b$, ∞, show that $a/\bar{d} = b/\bar{c} = c/\bar{b} = d/\bar{a}$, and deduce that $d = e^{i\theta}\bar{a}$ and $c = e^{i\theta}\bar{b}$ for some θ. By a judicious choice of multiple, show that every transformation fixing the unit circle $|z| = 1$ may be expressed in the form

$$z \mapsto \frac{az + b}{\bar{b}z + \bar{a}},$$

and that every transformation of this form fixes the unit circle.

53 Show that a transformation fixing the unit circle $|z| = 1$ as in qn 52, fixes the interior of the circle and if and only if $a\bar{a} - b\bar{b} > 0$.

Summary

Definition The set $\mathbb{C} \cup \{\infty\}$ is called the *Möbius plane*.
    qn 13
Definition A *Möbius transformation* is a transformation of the
    qn 13   Möbius plane of the form

$$z \mapsto \frac{az + b}{cz + d}$$

where $ad - bc \neq 0$, and $\infty \mapsto \infty$ when $c = 0$, and $-d/c \mapsto \infty \mapsto a/c$ otherwise.

Definition The *cross-ratio* $R(z_1, z_2; z_3, z_4)$ of four distinct points of
qns 11, 15 the Möbius plane is

$$\frac{z_3 - z_1}{z_3 - z_2} \Bigg/ \frac{z_4 - z_1}{z_4 - z_2}$$

with appropriate adjustments when one point is ∞.

Theorem The Möbius group is the full group of cross-ratio-
qns 15, 16 preserving transformations of the Möbius plane.

Theorem A Möbius transformation is uniquely determined by the
qns 17, 45 images of three points.

Theorem Four points of the Möbius plane are concyclic or col-
qns 21–24 linear if and only if their cross-ratio is real.

Theorem The set of circles and lines is transformed to itself by a
qns 25, 26 Möbius transformation if we adopt the convention of
adjoining ∞ to each line.

Definition If ON is a diameter of the sphere Σ and Π is the tan-
qn 35 gent plane at O, then the projection from N of
$\Sigma - \{N\}$ onto Π is called *stereographic projection*.

Definition The transformation of the Möbius plane $z \mapsto 1/\bar{z}$ which
qn 36 transposes 0 and ∞ is called the *inversion* in the unit
circle $|z| = 1$.

Theorem If two points are transposed by the inversion in the unit
qns 34, 37 circle, then any circle through those two points is
mapped onto itself by this inversion.

Theorem The (absolute value of the) angle between two circles is
qns 39–41 equal to the angle between their images under $z \mapsto 1/\bar{z}$.

Theorem The angle between two circles is equal to the angle
qn 42 between their images under any Möbius transformation.

Theorem The subgroup of the Möbius group fixing the upper half
qn 47 plane contains all transformations of the form

$$z \mapsto \frac{az + b}{cz + d},$$

where a, b, c, d are real and $ad - bc > 0$, and no
others.

Theorem The subgroup of the Möbius group fixing the interior of
qn 53 the unit disc contains all transformations of the form

$$z \mapsto \frac{az + b}{\bar{b}z + \bar{a}},$$

where $a\bar{a} - b\bar{b} > 0$, and no others.

Historical note

The first use of the equation

$$z' = \frac{az + b}{cz + d}$$

was by L. Euler (1777). It was B. Riemann (1854) who first adjoined '∞' to a plane to make $z \mapsto 1/z$ a continuous transformation. The notion of cross-ratio-preserving transformations is due to M. Chasles (1837), who was considering the cross-ratio of lengths of real line segments. The first use of complex numbers to describe projective transformations was by von Staudt (1856).

The systematic study of circle-preserving transformations was pursued by A.F. Möbius (1852–6) using synthetic methods. The interpretation of Möbius' results using complex numbers and the transformation

$$z \mapsto \frac{az + b}{cz + d}$$

is due to F. Klein (1875) and others.

Stereographic projection was used by Ptolemy (A.D. 140): its use for labelling the points of the sphere with $\mathbb{C} \cup \{\infty\}$ and its use in associating transformations of the sphere with transformations of the plane is due to F. Klein and is described in his *Lectures on the Icosahedron* (1884).

The subgroup of the Möbius group fixing the upper half plane is the group of direct isometries of the non-Euclidean plane with lines taken as semicircles centred on the real line. The subgroup of the Möbius group fixing the interior of the unit circle is the group of direct isometries of the non-Euclidean plane with lines taken as circular arcs orthogonal to the unit circle. Both these models of the hyperbolic plane are due to H. Poincaré (1881). The first major text book covering the whole of this area is by R. Fricke and F. Klein (1897).

Answers to chapter 4

1 π is a bijection.

2 $ONA = NAB = NOB$.

3 $\tan \theta$.

4 $ONB = \frac{1}{2}\pi - ONA$, so $OB' = \cot \theta$ and $OA' \cdot OB' = 1$.

5 B is labelled $1/x$. $x \mapsto 1/x$.

6 $\tan \theta \mapsto \tan (\theta + \frac{1}{2}\alpha)$ is

$$x \mapsto \frac{x + a}{1 - xa}.$$

$1/a \mapsto \infty \mapsto - 1/a$.

7 $\tan \theta \mapsto - \tan (\theta - \alpha)$ is

$$x \mapsto \frac{a - x}{1 + ax}.$$

$- 1/a \mapsto \infty \mapsto - 1/a$.

8 $c(ax + b) = a(cx + d) \Leftrightarrow ad - bc = 0$.

$$x = \frac{dr - b}{- cr + a}.$$

$- d/c$ has no real image. $- d/c \mapsto \infty \mapsto a/c$.

9 $\infty \mapsto \infty$.

10 Provided $ad - bc \neq 0$, all four are permutations of $\mathbb{R} \cup \{\infty\}$.

11 T is a subgroup by the argument of qn 3.1.
Thus a transformation of $\mathbb{R} \cup \{\infty\}$ preserves cross-ratio if and only if it has the form

$$x \mapsto \frac{ax + b}{cx + d}.$$

12 $\gamma = \alpha^{-1}\beta$.

13 (i) $\infty \mapsto \infty$. (ii) $- d/c \mapsto \infty \mapsto a/c$.

14 All but

$$z \mapsto \frac{bc - ad}{c} z$$

which requires $bc - ad \neq 0$.
First and third are direct similarities, the fourth is a translation.

15 Yes.

17 Use the algebra of qn 11.

18 The surface of the sphere with ∞ at N.

19 The circle centre $\frac{1}{2}$, radius $\frac{1}{2}$.
Transformation $z \mapsto 1/z$ transposes points so circle is mapped to line.

20 z and c lie on a circle through w and t.

21 0 or π so the number is real.

22 0 or π.

23 0 or π.

24 (i) The circle through z, t and w. (ii) The line through z, t, and w, and the point ∞.

25 Since Möbius transformations preserve cross-ratio, they preserve real cross-ratios, so the image of a circle is either a circle or a line with ∞.

26 As in qn 25 the image is either a circle or a line with ∞.

27 If they are parallel, a translation is sufficient. If they intersect, a rotation is sufficient. Every direct isometry is a Möbius transformation.

28 Choose three points a, b, c on the circle and three points a', b', c' on the line, then by qns 12 and 17 there is a Möbius transformation mapping $a \mapsto a'$, $b \mapsto b'$, $c \mapsto c'$ and from qn 25, the circle maps onto the line with ∞.

29 $e^{i\theta} \mapsto e^{-i\theta}$. $2 \mapsto \frac{1}{2}$. $|z| = 2$ maps to $|z| = \frac{1}{2}$. $\mathbb{R} - \{0\} \to \mathbb{R} - \{0\}$.
The line is reflected about the real axis, but not pointwise.

30 $\{r, 1/r, e^{i\theta}\} \to \{1/r, r, e^{-i\theta}\}$. If the line joining 0 to $e^{i\theta}$ meets the circle again at z, $1 \cdot |z| = r \cdot (1/r)$ so $|z| = 1$, and $z = e^{i\theta}$. C' is the reflection of C in the real axis. C' cuts $|z| = 1$ orthogonally.

33 $rs = 1$. r and s are transposed. Σ is mapped to its reflection in the real axis.

34 $\theta = \phi$, $rs = 1$. $re^{i\theta} \cdot \overline{se^{i\theta}} = 1$.

35 $z \mapsto 1/\bar{z}$. In each plane through ON the reflection acts like the real transformation $x \mapsto 1/x$.

36 $z \mapsto 1/\bar{z}$ maps circles and lines to circle and lines.

37 Σ is mapped to itself, but not pointwise.

38 Any circle through P and P' is orthogonal to $|z| = 1$, so any such circle is mapped to itself under $z \mapsto 1/\bar{z}$. So $S \to S$. Thus $S \cap C \to S \cap C'$. Since $S \cap C$ is a single point, so is $S \cap C'$.

39 C_1 and S_1 have a common tangent at P, C_2 and S_2 have a common tangent at P so angle between C_1 and C_2 = angle between S_1 and S_2.

40 (i) Argument of qns 38 and 39 holds so long as $P \neq P'$ and $P \neq 0$. (ii) Circle image is parallel to tangent at 0.

41 Apply argument of qns 38 and 39 to the point of intersection off $|z| = 1$.

42 $z \mapsto cz + d, z \mapsto \bar{z}, z \mapsto 1/\bar{z},$

$$z \mapsto \frac{bc - ad}{c} z,$$

$z \mapsto z + a/c$ each preserve angles of intersection and their product is

$$z \mapsto \frac{az + b}{cz + d}.$$

43 (i) $z \mapsto az + b,$ 　　　　　(v) $z \mapsto az + 1 - a,$

(ii) $z \mapsto \dfrac{az}{cz + d},$ 　　　　(vi) $z \mapsto \dfrac{az}{cz + a - c}.$

(iii) $z \mapsto \dfrac{az + c + d - a}{cz + d},$

(iv) $z \mapsto az,$

The set of all transformations of the given form form the stabiliser in each case.

44 If z is a fixed point,

$$z = \frac{az + b}{cz + d}$$

so $cz^2 + (d - a)z - b = 0$.
This quadratic equation has either 1 or 2 roots. If $c = 0$, then ∞ and $b/(d - a)$ are fixed.

45 Only the identity fixes three points. If two transformations α, β map three points in the same way, then $\alpha\beta^{-1}$ fixes the three points and $\alpha\beta^{-1} = 1$. $a \mapsto a', b \mapsto b', c \mapsto c'.$

46 From qn 44, equal roots when $(d - a)^2 + 4bc = 0$.

47 $z \mapsto az + b$ fixes ∞, and fixes no other point if $a = 1$.

48 If real line is fixed $0 \mapsto a_1, 1 \mapsto a_2, \infty \mapsto a_3$, where a_1, a_2, a_3 are real from qns 11 and 17.

49 $z \mapsto \dfrac{az + b}{cz + d}$

preserves upper half plane if Im $(z) > 0$ implies

$$\text{Im}\left(\frac{az + b}{cz + d}\right) > 0.$$

50 0, w, $1/\bar{w}$ are collinear. Product of distances from $0 = 1$. Every circle through w and $1/\bar{w}$ cuts $|z| = 1$ orthogonally so by angle preservation every circle through $w\alpha$ and $(1/\bar{w})\alpha$ cuts $|z| = 1$ orthogonally. The line through $w\alpha$ and $(1/\bar{w})\alpha$ cuts $|z| = 1$ orthogonally and so passes through 0. $\gamma: z \mapsto 1/\bar{z}$, $w\alpha\gamma = w\gamma\alpha$.

51 With γ as in qn 50, $0\alpha = 0$ so $\infty\alpha = 0\gamma\alpha = 0\alpha\gamma = 0\gamma = \infty$. $0 \mapsto 0$, $\infty \mapsto \infty$ and $1 \mapsto e^{i\theta}$.

52 With γ as in qn 50, $z\alpha\gamma = z\gamma\alpha$. $d\bar{d} = a\bar{a}$ implies $|d| = |a|$ so $|a/\bar{d}| = 1$ and $d/\bar{a} = e^{i\theta}$ (say).
$$\frac{az + b}{cz + d} = \frac{az + b}{\bar{b}e^{i\theta}z + \bar{a}e^{i\theta}}.$$
Now multiply top and bottom by $e^{-\frac{1}{2}i\theta}$.

53 $0 \mapsto b/\bar{a}$. $|b/\bar{a}| < 1 \Leftrightarrow |b| < |\bar{a}| \Leftrightarrow a\bar{a} - b\bar{b} > 0$.
If w is interior, there is a circle through 0 and w not intersecting $|z| = 1$. This circle's image cannot intersect $|z| = 1$.

5

The regular solids

In chapter 3, we found the group of isometries of the real plane and some of its subgroups. In this chapter we will identify three finite groups of isometries of real 3-dimensional space. These will be the groups of rotational symmetries of the regular tetrahedron, the cube and the icosahedron respectively. As regular solids, these may be inscribed in a sphere, and then any symmetry of any one of the solids will leave the centre of the sphere fixed and will transform the surface of the sphere onto itself. As we will prove in chapter 20, rotations are the only direct isometries of 3-dimensional space which fix a point.

It is important to make the models suggested in qns 1, 2 and 3 and to keep the models to hand as you explore their rotational symmetries.

Concurrent reading: Coxeter (1969), p. 151; Steinhaus, pp. 208f, 216f; Hilbert and Cohn-Vossen, pp. 89 – 93;Martin, chapter 17; Klein, chapter 1.

1 A *regular solid* is a 3-dimensional polyhedron in which each face is a regular polygon. Any two faces may be matched by an isometry and any two vertices may also be matched by an isometry.

If enough of the edges are cut to allow the faces of the polyhedron to lie, connected, in a plane, then the outline so formed is called the *net* of the solid. By examining the possible nets, determine whether a regular solid with each face an equilateral triangle may possibly be constructed with 2, 3, 4, 5, 6 or more faces meeting at a vertex.

Construct the regular solids, with equilateral triangles as faces, which you have not ruled out from theoretical considerations. (The tetrahedron, octahedron and icosahedron.)

2 Determine whether a regular solid with each face a square may possibly be constructed with 2, 3, 4 or more faces meeting at a vertex.
Construct the one regular solid with square faces which is not ruled out from theoretical considerations. (The cube.)

3 Determine whether a regular solid with each face a regular pentagon may possibly be constructed with 2, 3, 4 or more faces meeting at a vertex.
Construct the one regular solid with pentagonal faces which is not ruled out from theoretical considerations. (The dodecahedron.)

4 Determine whether a regular solid may be constructed with hexagonal faces.

5 What figure is formed by taking as vertices the midpoints of the faces of a (i) tetrahedron, (ii) cube, (iii) octahedron, (iv) dodecahedron, (v) icosahedron?

6 By counting how many rotational symmetries of a regular tetrahedron stabilise one vertex, and also counting the number of distinct images that vertex has under the group of rotational symmetries of the tetrahedron, show that the number of rotational symmetries of a regular tetrahedron cannot exceed 12.

7 By labelling the faces of a regular tetrahedron 1, 2, 3, 4 (or, if you prefer, labelling the vertices), describe each rotational symmetry of the tetrahedron that you can find as a permutation in S_4. Can you find 12 such? Do they form a familiar subgroup of S_4?

8 Use the labelling of qn 7 to describe reflection symmetries of the tetrahedron. Are there other elements of S_4 which correspond to symmetries of the tetrahedron other than the rotations and reflections?

9 By counting how many rotational symmetries of a cube stabilise one vertex, and also counting the number of distinct images of that vertex show that the number of rotational symmetries of a cube cannot exceed 24. Do you get the same number if you count the rotational symmetries stabilising a face and the number of images of a face?

10 Describe each rotational symmetry of the cube that you can find as an element of S_4 by labelling pairs of opposite vertices of a cube (or the diagonals of a cube) with the digits 1, 2, 3 and 4. Are there 24 rotational symmetries of a cube?

11 By labelling pairs of opposite faces of a regular octahedron 1, 2, 3, 4 describe the rotational symmetries of the octahedron as elements of S_4 and state the angle of the corresponding rotation in each case.

12 Use qn 5 to explain the connection between the results of qns 10 and 11.

13 By counting how many rotational symmetries of a regular icosa-hedron stabilise one vertex, and also counting the number of the distinct images that a vertex has under the group of rotational symmetries of the icosahedron, show that the number of rotational symmetries cannot exceed 60. Do you get the same number if you count the rotational symmetries stabilising a face and the number of different images of a face?

14 If the faces of an icosahedron are labelled as shown, which elements of S_5 correspond to rotational symmetries of the icosahedron through $\pm 72°$ (12), through $\pm 144°$ (12), through $\pm 120°$ (20), through $180°$ (15)? Are these permutations even? Do they form a group?

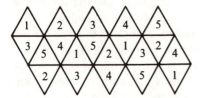

15 What can be said about the rotational symmetries of a dodecahedron?

That the finite sets of rotations we have identified do indeed form groups will follow when we have proved that the product of two rotations in three dimensions, with a common fixed point, is again a rotation. To do this, we use the 3-dimensional analogue of qn 3.21. In qn 3.21 we showed that if two lines in a plane are inclined at an angle of $\frac{1}{2}\theta$ and meet at the point O, then a rotation of the plane about the point O, through an angle θ may be expressed as the product of the reflections in each of the two lines.

16 Let OA, OB and OC be lines in three dimensions in different direc-tions, and let A, B and C be points on the surface of a sphere with centre O. Let α denote the reflection in the plane OBC, let β denote the reflection in the plane OCA and let γ denote the reflec-tion in the plane OAB. Interpret the equation $(\gamma\alpha)(\alpha\beta) = \gamma\beta$. If rotations about the axes OB and OC are given, explain how to find the point A so that the given equation determines their product.

Summary

Theorem There are five regular solids.
qns 1–4

Theorem The group of rotational symmetries of a regular tetra-
qn 7 hedron is A_4.

qn 8 The full group of symmetries of a regular tetrahedron is
S_4.

Theorem The group of rotational symmetries of a cube or of a
qns 10, 11 regular octahedron is S_4.

Theorem The group of rotational symmetries of a regular
qns 14, 15 icosahedron or of a regular dodecahedron is A_5.

Historical note

The uniqueness of the five regular solids and the length of their
circumradii are established in *Euclid Book XIII* (300 B.C.). The later
books *Euclid XIV* and *XV* (A.D. 300) establish the dual nature of the
cube and octahedron on the one hand and of the icosahedron and
dodecahedron on the other. The symmetry groups of these solids were
described by J. F. C. Hessel (1830) and A. Bravais (1849) in their
classifications of crystal structures. The isomorphisms of their groups
of rotations with the permutation groups A_4, S_4 and A_5 were known to
F. Klein by 1874 and were expounded in the first chapter of his
Lectures on the Icosahedron (1884).

Answers to chapter 5

1 No solid can have only two faces meeting at a vertex. $n \cdot 60° < 360° \Rightarrow n = 3, 4, 5$.

2 $n \cdot 90° < 360° \Rightarrow n = 3$.

3 $n \cdot 108° < 360° \Rightarrow n = 3$.

4 $n \cdot 120° < 360° \Rightarrow n < 3$, contradiction.

5 (i) tetrahedron, (ii) octahedron, (iii) cube, (iv) icosahedron, (v) dodecahedron.

6 Three rotations stabilise one vertex. Each vertex has four images. At most $3 \cdot 4 = 12$ rotations in all.

7 A_4.

8 The six transpositions give the reflections. The 4-cycles give neither reflections nor rotations.

9 Three rotations stabilise one vertex. Each vertex has eight images. At most $3 \cdot 8 = 24$ rotations in all. Four rotations stabilise one face. Each face has six images. $4 \cdot 6 = 24$.

10 Yes.

11 4-cycles, 90°. 3-cycles, 120°. Transpositions or pairs, 180°.

12 Every rotational symmetry of a cube is a rotational symmetry of its inscribed octahedron and vice-versa.

13 Five rotations stabilise one vertex. Each vertex has 12 images. Three rotations stabilise one face. Each face has 20 images.

14 $\pm 72°$: (12345), (15432), (12534), (14352), (13425), (15243), (13254), (14523), (14235), (15324), (12453), (13542). $\pm 144°$: (13524), (14253), (15423), (13245), (14532), (12354), (12435), (15342), (12543), (13452), (14325), (15234). $\pm 120°$: all 3-cycles. $\pm 180°$: all pairs of transpositions. A_5.

15 Can be identified through qn 5 with the rotational symmetries of an icosahedron.

16 Argue as in qn 3.24. $\gamma\alpha$ is a rotation about OB; $\alpha\beta$ is a rotation about OC; $\gamma\beta$ is a rotation about OA. Locate A so that the angle between the planes OAB and OBC is half the angle of the rotation about OB, and so that the angle between the planes OBC and OAC is half the angle of the rotation about OC.

6

Abstract groups

We start this chapter by abstracting the properties of S_4 to which we drew attention in qn 1.34 and use these properties to define a group in a way which does not depend on permutations or geometry and in which the elements of the group need not even be functions. For two functions, composition has been presumed to be the method of combination. For two numbers, addition or multiplication are possible methods of combination. In general the phrase 'binary operation' is used to describe the combination of two elements to make one element. Some immediate consequences of the group axioms are identified in qns 5–9.

Concurrent reading: Green, chapters 4 and 5, Fraleigh, sections 2, 3, 6 and 7.

Axioms

So far every group we have considered has been a subgroup of a symmetric group. We now widen the definition of a *group* to include those sets, G, with a binary operation ($\cdot$), not just composition of functions, which satisfy the four conditions which we insisted on for symmetric groups, namely

closure If $a, b \in G$, then $a \cdot b \in G$.

associativity If $a, b, c \in G$, then $(a \cdot b) \cdot c = a \cdot (b \cdot c)$.

identity There exists $e \in G$ such that $a \cdot e = e \cdot a = a$ for all $a \in G$.

inverses For each $a \in G$, there exists an a^{-1} such that $a \cdot a^{-1} = a^{-1} \cdot a = e$.

1 Name some sets of numbers which under the binary operation of addition form groups according to the definition above.

2 Name some sets of numbers which under the binary operation of multiplication form groups according to the definition above.

3 For what numbers does $a - (b - c) = (a - b) - c$? Do any of the sets you have chosen for qn 1 form groups under subtraction?

4 For what numbers does $a/(b/c) = (a/b)/c$? Do any of the sets you have chosen for qn 2 form groups under division?

5 If e and f are elements of a group $(G, \cdot)$ and for every element a in G, $a \cdot e = e \cdot a = a = a \cdot f = f \cdot a$, prove that $e = f$. [*The identity is unique.*]

6 If $(a^l) \cdot a = e$, a^l is called a *left inverse* for a. If $a \cdot (a^r) = e$, a^r is called a *right inverse* for a. By considering the product $a^l \cdot a \cdot a^r$ prove that every left inverse is equal to every right inverse and hence that the inverse of a is unique.

7 If $(G, \cdot)$ is a group with given elements a and b, prove that there is a unique x in G such that $a \cdot x = b$, and a unique y in G such that $y \cdot a = b$. What is the inverse of a^{-1}?

8 Show that every group $(G, \cdot)$ contains a unique element a such that $a^2 = a \cdot a = a$.

9 If a and b are elements of a group $(G, \cdot)$, find $(ab)(b^{-1}a^{-1})$, and deduce that $(ab)^{-1} = b^{-1}a^{-1}$. Give an example to show that $(ab)^{-1} = a^{-1}b^{-1}$ may be false.

Subgroups

When a subset H of a group $(G, \cdot)$ has the property that $(H, \cdot)$ is itself a group, then H is called a *subgroup* of G. When $G \neq H$, H is called a *proper* subgroup. The smallest subgroup containing a given element a is denoted by $\langle a \rangle$ and is said to be *generated* by a. Such subgroups, generated by a single element, are useful in analysing the structure of groups and are said to be *cyclic*. Finite cyclic groups lead to the notion of the *order* of an element. For the element a, the order is the least positive integer n such that $a^n = e$.

10 Name some subgroups of $(\mathbb{C}, +)$ and some subgroups of $(\mathbb{C} - \{0\}, \times)$.

11 If H is a subgroup of G, use qn 8 to show that these two groups have the same identity. Deduce that the inverse of an element in H is the same as its inverse in G.

12 If H and K are both subgroups of G, prove that $H \cap K$ is also a subgroup of G.

Give examples from the groups D_2 and S_3 to show that $H \cup K$ need not be a subgroup of G.

13 If the group $(\mathbb{R}, +)$ contains a subgroup $(H, +)$ and 1 is in H, what other elements of $\mathbb{R}$ can we be sure lie in H? Must every integer lie in H? (In the light of this question $(\mathbb{Z}, +)$ is said to be the subgroup of $(\mathbb{R}, +)$ *generated* by 1, and this is expressed symbolically by writing $\langle 1 \rangle = (\mathbb{Z}, +)$.)

14 If $(\mathbb{Z}, +)$ contains a subgroup H and 2 is in H, what other elements of $\mathbb{Z}$ can we be sure must lie in H? (The subgroup generated by 2, $\langle 2 \rangle$, in $(\mathbb{Z}, +)$ consists of the even integers.)

15 If $\alpha = (1234)$ and $\beta = (12)(34)$ denote symmetries of a square described by permutations of its vertices, find α^2, α^3, $\beta\alpha$, $\beta\alpha^2$ and $\beta\alpha^3$, as permutations of the vertices, and describe them geometrically as symmetries of the square. This is an example of the dihedral group D_4. What is the smallest subgroup containing α, that is, $\langle \alpha \rangle$, the subgroup generated by α? Find the subgroups $\langle \alpha^2 \rangle$, $\langle \alpha^3 \rangle$, $\langle \beta \rangle$, $\langle \beta\alpha \rangle$, $\langle \beta\alpha^2 \rangle$, $\langle \beta\alpha^3 \rangle$ and $\langle I \rangle$ respectively.

Order

16 If α is a reflection of the Euclidean plane, say, $\alpha : z \mapsto \bar{z}$, then $\alpha^2 = I$, the identity function, and α is said to have order 2. The reflection α generates the subgroup $\{I, \alpha\}$ of $S_\mathbb{C}$.

If β is a rotation of the Euclidean plane through 120°, say, $\beta : z \mapsto e^{2\pi i/3}z$, find the least positive power n of β such that $\beta^n = I$ (n is then called the *order* of β) and identify the subgroup of $S_\mathbb{C}$ which is generated by β.

17 Make a list of elements of S_3 and state the order of each element. What is the subgroup of S_3 generated by (123)?

18 State the orders of the elements of D_4 as given in qn 15.

Cyclic groups: groups generated by one element

19 For any element g in a group $(G, \cdot)$, we define $g^0 = e$, $g^1 = g$, and then $g^{n+1} = (g^n)g$ for positive integers n as in qn 2.7. Also we define $g^{-n} = (g^{-1})^n$ as in qn 2.7 for positive integers n. Assuming that the results of qn 2.7 apply in this more general context, show that the set $\{g^r | r = 0, \pm 1, \pm 2, \dots\}$ is a subgroup of G. Must any subgroup of G which contains the element g contain the whole of this set? As before, we denote this subgroup by $\langle g \rangle$.

20 A subgroup of the form $\langle g \rangle$ in any group, that is to say, a *subgroup generated by one element* is known as a *cyclic subgroup*.
Find a cyclic subgroup of S_5 containing exactly five elements.
Find a cyclic subgroup of S_6 containing exactly six elements.

21 If a is a group element of order 5, how many distinct elements are there in $\langle a \rangle$? What are the orders of each of these elements?

22 If a is a group element of order 6, how many distinct elements are there in $\langle a \rangle$? What are the orders of each of these elements?

23 If a is a group element of order 6, find all the subgroups of $\langle a \rangle$. Are all these subgroups cyclic?

24 If a and b are elements of a cyclic group, must it follow that $ab = ba$?

25 If a is an element of a group and a has order n, what can be said about the order of a^{-1}?

26 The group $(\mathbb{Z}, +)$ is a cyclic group generated by 1. If $(H, +)$ is a subgroup of $(\mathbb{Z}, +)$ and the least positive integer in H is a, prove that H can only contain the elements of $\langle a \rangle$, and deduce that every subgroup of $(\mathbb{Z}, +)$ is cyclic.

27 If G is an infinite cyclic group, prove that every subgroup of G is cyclic.

28 If G is a cyclic group generated by an element g of order n, we know from qn 19 that $G = \langle g \rangle = \{g^r | r \in \mathbb{Z}\}$.
Prove that $G = \langle g \rangle = \{g^r | 0 \leqslant r < n\}$, and that the n elements in this set are distinct. Let H be a subgroup of G and of all the elements of H let g^a be that with least positive exponent. Prove that if k is a positive integer less than a, then no element of the form g^{la+k} may belong to H. Deduce that $H = \langle g^a \rangle$.

Groups generated by two elements

If a group contains the elements a and b, then it clearly contains both $\langle a \rangle$ and $\langle b \rangle$. What other elements it must contain depends on the relation between the elements a and b. The smallest subgroup containing a and b is called the *subgroup generated by a and b* and is denoted by $\langle a, b \rangle$. The simplest relation between a and b is the equation $ab = ba$, as, for example, when a and b are disjoint cycles of a permutation group. When such an equation holds, we say that a and b *commute*, or satisfy the commutative law.

29 If a and b are elements of a group and $ab = ba$, prove by induction on n that $a^n b = ba^n$.

If, in addition, the element b is given to have order 2, by filling in the multiplication table shown,

	a^i	ba^i
a^m		
ba^n		

prove that the set $\{a^n, ba^n | n \in \mathbb{Z}\}$ is closed, and then complete the proof that it forms a subgroup, the subgroup $\langle a, b \rangle$.

30 If a and b are elements of a group and $ab = ba$, prove that $a^n b^2 = b^2 a^n$. (It is only necessary to use qn 29, no induction is required.)

If, in addition, b has order 3, by filling in the multiplication table shown, prove that the set

	a^i	ba^j	$b^2 a^k$
a^l			
ba^m			
$b^2 a^n$			

$\{a^n, ba^n, b^2 a^n | n \in \mathbb{Z}\}$ is closed, and then complete the proof that it forms a subgroup, the subgroup $\langle a, b \rangle$.

31 If a and b are elements of a group and $ab = ba$, prove that $a^m b^n = b^n a^m$ for all integers m and n. (Use qn 29 and an induction on n.) Deduce that $\{a^m b^n | m, n \in \mathbb{Z}\}$ forms a subgroup, the subgroup $\langle a, b \rangle$.

32 If a and b are elements of a group and $(ab)^2 = a^2 b^2$, prove that $ab = ba$.

33 If a and b are elements of a group and $ab = ba$, prove that $(ab)^n = a^n b^n$ for all integers n.

34 If $a = (12)$ and $b = (345)$, what is the order of ab?

35 If $a = (12)$ and $b = (34567)$, what is the order of ab?

36 If $a = (1234)$ and $b = (567890)$, what is the order of ab?

37 If a and b are disjoint cycles of length m and n respectively, and the highest common factor of m and n is k, what is the order of ab?

38 Give an example to show that if a and b are elements of a group and a has order 6 and b has order 10, then ab need not have order 30 even when $ab = ba$.

When $ab = ba$ for any two elements a, b of a group $(G, \cdot)$, then $(G, \cdot)$ is said to be *abelian*.

Dihedral groups

Another relationship that may hold between two elements a and b of a group is $ab = ba^{-1}$ and in this case too it is possible to give a complete account of the elements of the group $\langle a, b \rangle$. We will concentrate only on such groups for which b has order 2.

39 If $a = (123)$, $b = (23)$, so that $a^3 = b^2 = e$, verify that $ab = ba^{-1}$ and, by listing the permutations a^2, ba and ba^2, verify that $\langle a, b \rangle = S_3$.

40 If $a = (1234)$ and $b = (12)(34)$, so that $a^4 = b^2 = e$, verify that $ab = ba^{-1}$ and, by listing the permutations a^2, a^3, ba, ba^2 and ba^3, verify that $\langle a, b \rangle$ describes the group of symmetries of a square as in qn 15.

41 If $a = (12345)$ and $b = (25)(34)$, so that $a^5 = b^2 = e$, verify that $ab = ba^{-1}$. List the permutation a^2, a^3, a^4, ba, ba^2, ba^3 and ba^4 and, by labelling vertices of a regular pentagon appropriately with the digits 1, 2, 3, 4 and 5, name geometric symmetries which correspond to these permutations.

42 If a and b are elements of a group, b has order 2 and $ab = ba^{-1}$, prove that $bab = a^{-1}$. Evaluate the product $(bab)(bab)$ in two ways to establish that $ba^2b = a^{-2}$ and hence that $a^2b = ba^{-2}$. Extend this method to establish that $a^n b = ba^{-n}$ for all integers n. By filling in the multiplication table shown,

	a^k	ba^l
a^m		
ba^n		

prove that the set $\{a^n, ba^n \mid n \in \mathbb{Z}\}$ is closed, and then complete the proof that it forms a subgroup, the subgroup $\langle a, b \rangle$.

43 A group G is generated by an element a of order n and an element b of order 2 with the relation $ab = ba^{-1}$. Prove that G contains $2n$ elements and that every element of the form ba^r has order 2. This

is the abstract definition of the *dihedral group* D_n. The name
dihedral (two-faced) stems from the fact that in the plane, the
group of symmetries of a regular n-gon is of this type. See
qns 3.41–47.

Groups generated by larger sets of elements

44 Express each element of S_3 as a product of transpositions. This
possibility may be expressed by writing $\langle (12), (13), (23) \rangle = S_3$.

45 Use qn 2.20 to show that the set of transpositions in S_n generates S_n.

46 Evaluate $(1a)(1b)(1a)$ and deduce that the set of transpositions
$\{(12), (13), \ldots , (1n)\}$ generates S_n.

47 Evaluate $(12)(13)$, and also evaluate $(134)(132)$ as a product of disjoint cycles, and deduce that every even permutation may be
expressed as a product of 3-cycles. That is, the 3-cycles generate
A_n.

48 Explain how to express a translation as a product of two reflections
and how to express a rotation as a product of two reflections. This
establishes that every transformation of the form $z \mapsto e^{i\theta}z + c$ is
the product of two reflections. Deduce that every transformation of
the form $z \mapsto e^{i\theta}\bar{z} + c$ is a product of at most three reflections,
and that the set of reflections generates the Euclidean group.

49 Since every element in the group of direct isometries is a product of
two reflections, deduce that every product of four reflections or of
any product of an even number of reflections is equal to a product
of two reflections. Prove also that the product of any odd number
of reflections is either equal to a single reflection or to a product of
three reflections.

Isomorphism

When can we say that two groups are the same? In some cases this
may be obvious. For example, if the elements and the relation between
those elements are written down using roman letters, we have the same
group as if they had been written down using Greek letters. In qns 2.49
and 3.47 and qn 40, we have found different ways of representing the
same group of symmetries. We now examine this kind of possibility.

50 The group of symmetries of a rectangle, D_2, may be exhibited as a
subgroup of S_4 by labelling the vertices of the rectangle with the
numbers 1, 2, 3 and 4. See qn 2.53.
What is geometrically the same group may be exhibited as a

different subgroup of S_4 by labelling the vertices of a rhombus and
exhibiting its group of symmetries. See qn 2.46.
Again, each of the elements of this geometrical group may be
expressed as an isometry in S_C, or a mapping of coordinate pairs
(x, y). See qn 3.41.
Exhibit these four presentations of this geometric group consisting
of the identity, two reflections and a half-turn.

51 Exhibit the group of symmetries of an equilateral triangle, D_3
   (i) as a subgroup of S_3 by labelling the vertices 1, 2, 3;
   (ii) as a subgroup of S_6 by labelling the points one-third of the way
      along each of the sides.
Give the two lists of elements in such a way that it is plain which
elements of the two groups match.

52 When there is a one–one matching of the elements of a group G with
   the elements of a group G' with g in G matching g' in G', then the
   function $G \to G'$ given by $g \mapsto g'$ is called an *isomorphism* and the
   groups G and G' are said to be *isomorphic*, provded that
   $(ab)' = a'b'$, for every pair of elements a, b in G.
   We then write $G \cong G'$.
   If $a = (123)$ and $\alpha{:}z \to e^{\frac{2}{3}\pi i}z$ exhibit multiplication tables for each of
   the groups $\{e, a, a^2\}$ and $\{I, \alpha, \alpha^2\}$ and say how the elements
   should be matched for an isomorphism between these two groups.

53 By using the convention that the product ab appears in the row with
   a at the left and b at the top, a binary operation on a finite set
   may be exhibited by means of a multiplication table.

	b
a	ab

For the table given here, say how you would find the identity, check
for closure, and then find an inverse for each element.

	a	b	c	d
a	a	b	c	d
b	b	a	d	c
c	c	d	a	b
d	d	c	b	a

Use qn 7 to explain why the elements in a row (or a column) of a group
table are all different.

54 Because the multiplication table or *Cayley table* for a group exhibits products, the matching of two Cayley tables is a check on the structure-preserving property $(ab)' = a'b'$.
Which of the following Cayley tables for a group may be rearranged so as to establish an isomorphism with the group exhibited in qn 53?

	p	*I*	*q*	*r*
p	*I*	*p*	*r*	*q*
I	*p*	*I*	*q*	*r*
q	*r*	*q*	*I*	*p*
r	*q*	*r*	*p*	*I*

	I	*l*	*m*	*n*
I	*I*	*l*	*m*	*n*
l	*l*	*m*	*n*	*I*
m	*m*	*n*	*I*	*l*
n	*n*	*I*	*l*	*n*

55 The Cayley table

·	1	2	3
1	1	2	3
2	2	3	1
3	3	1	2

is that of a group, *G*.
Make a Cayley table for the permutation group {(1), (123), (132)} and check that *G* is isomorphic to A_3 under the matching $1 \mapsto (1)$, $2 \mapsto (123)$, $3 \mapsto (132)$. Exhibit the functions $x \mapsto x \cdot 1$, $x \mapsto x \cdot 2$ and $x \mapsto x \cdot 3$ of *G*. What can you say about the matching $a \mapsto [x \mapsto x \cdot a]$?

56 For each of the groups with Cayley tables

	1	2	3	4
1	1	2	3	4
2	2	3	4	1
3	3	4	1	2
4	4	1	2	3

	1	2	3	4
1	1	2	3	4
2	2	1	4	3
3	3	4	1	2
4	4	3	2	1

construct the four permutations $x \mapsto x \cdot 1$, $x \mapsto x \cdot 2$, $x \mapsto x \cdot 3$ and $x \mapsto x \cdot 4$. Is the matching $a \mapsto [x \mapsto x \cdot a]$ an isomorphism in each case?

57 For any group $(G, \cdot)$ show that the set of functions of the form $x \mapsto x \cdot a$ forms a subgroup of S_G. By matching the function

$x \mapsto x \cdot a$ with the element a show that this subgroup of S_G is isomorphic to $(G, \cdot)$.

It is this theorem (Cayley, 1878) which shows that our starting point in chapter 1 was sufficiently general to give an isomorphic image of every possible group.

58 Establish that the group of translations of $\mathbb{R}$ is isomorphic to $(\mathbb{R}, +)$.

59 Establish that the group of translations of $\mathbb{C}$ is isomorphic to $(\mathbb{C}, +)$.

60 In the group of ratio-preserving transformations of $\mathbb{R}$, prove that the stabiliser of 0 is isomorphic to $(\mathbb{R} - \{0\}, \times)$.

61 In the Möbius group, use the fact that the stabiliser of ∞ is the group of direct similarities to prove that the stabiliser of 0 and ∞ is isomorphic to $(\mathbb{C} - \{0\}, \times)$.

62 Use the logarithmic function to establish that the multiplicative group of positive real numbers is isomorphic to $(\mathbb{R}, +)$.

63 If a cyclic group contains n distinct elements, what can be said about the order of one of its generators?

64 If two cyclic groups each contain n distinct elements, establish that they are isomorphic. The symbol C_n is used to describe such groups up to isomorphism.

65 Show that two infinite cyclic groups are isomorphic.

66 If $\alpha: G \to G'$ is an isomorphism and e is the identity in G and e' is the identity in G', prove that $e\alpha = e'$.

67 If $\alpha: G \to G'$ is an isomorphism and $g \in G$, prove that $(g^{-1})\alpha$ is the inverse of $g\alpha$ in G'. In other words, $(g^{-1})\alpha = (g\alpha)^{-1}$.

68 If $\alpha: G \to G'$ is an ismorphism and g is an element of order n in G, prove that $g\alpha$ is an element of order n in G'.

69 Determine whether either the full group of symmetries of a cuboid with three unequal sides or the full group of symmetries of a solid swastika is isomorphic to D_4.

Summary

Definition A set G on which a binary operation $\cdot$ has been
defined is called a *group* when

(i) $a \cdot b \in G$ for all a, $b \in G$ (*closure*),

(ii) $a \cdot (b \cdot c) = (a \cdot b) \cdot c$ for all a, b, $c \in G$ (*associativity*),

(iii) there exists $e \in G$ such that $a \cdot e = e \cdot a = a$ for all $a \in G$
(*identity*),

(iv) for each $a \in G$, there exists $a^{-1} \in G$ such that
$a \cdot a^{-1} = a^{-1} \cdot a = e$. (*inverses*).

Theorem The identity in a group is unique. The inverse of each
qns 5, 6, element is unique. Solutions for x of $a \cdot x = b$ and
7 $x \cdot a = b$ are each unique.

Definition If H is a subset of a group $(G, \cdot)$, and $(H, \cdot)$ is a group,
qn 10 then H is called a *subgroup* of G.

Theorem The intersection of two subgroups is a subgroup.
qn 12

Definition If $h \in G$, the smallest subgroup of G which contains h,
qn 13 or, the intersection of all subgroups of G which contain
h, forms a subgroup which is denoted by $\langle h \rangle$ and is
called the subgroup of G *generated* by h.

Definition A group or subgroup which is generated by a single
qn 20 element is said to be *cyclic*.

Theorem Every subgroup of a cyclic group is cyclic.
qns 27, 28

Definition A group generated by two elements a and b with b of
qn 43 order 2 is called a *dihedral* group if $ab = ba^{-1}$.

Definition If $\alpha : G \to G'$ is a bijection of groups, and for any
qn 52 a, $b \in G$, $(a \cdot b)\alpha = (a\alpha) \cdot (b\alpha)$, the groups G and G' are
said to be *isomorphic*, and α is called an *isomorphism*.

Theorem If $\alpha : G \to G'$ is an isomorphism and if e is the identity
qns 66–68 in G then $e\alpha$ is the identity in G'. If a^{-1} is the inverse of
a in G then $a^{-1}\alpha$ is the inverse of $a\alpha$ in G'. If a has
order n in G, then $a\alpha$ has order n in G'.

Historical note

E. Galois (1830), to whom we owe the term *group*, C. Jordan (1870)
and F. Klein (1884) used only the closure axiom to define a group. The
other axioms are implicit in their work because it was finite sets of permutations or transformations which were under consideration. In a
paper by A. Cayley (1854) the need for the associative law and for the
existence of an identity is explicit, and Cayley defined a group using
abstract symbols and either a multiplication table or a set of defining
relations. In 1882, both W. Dyck and H. Weber published modern

axioms for a group, and these became widely used after the publication of Weber's textbook in 1886.

A. L. Cauchy defined the order of an element of a group in 1815.

The need for such a notion as isomorphism is inherent in C. F. Gauss *Disquisitiones Arithmeticae* (1801) with the many types of abelian groups which he considered (additive and multiplicative groups modulo *n*, *n*th roots of unity, composition of quadratic forms with the same discriminant) and in fact the notion of isomorphism appears under various names in the work of most nineteenth-century group theorists. It became the definitive form of 'sameness' in the logical work of A. N. Whitehead and B. Russell (1910).

Answers to chapter 6

1 $\mathbb{Z}, \mathbb{Q}, \mathbb{R}, \mathbb{C}$.

2 $\mathbb{Q} - \{0\}, \mathbb{R} - \{0\}, \mathbb{C} - \{0\}$.

3 $a - (b - c) \neq (a - b) - c$ unless $c = 0$. So the singleton $\{0\}$ is the only group under subtraction.

4 $a/(b/c) \neq (a/b)/c$ unless $c = \pm 1$. So $\{\pm 1\}$ and $\{1\}$ are the only groups under division.

5 $e = e \cdot f = f$.

6 $a^1 = a^1(a \cdot a^r) = (a^1 \cdot a)a^r = a^r$.

7 $ax = b \Rightarrow a^{-1}(ax) = a^{-1}b \Rightarrow x = a^{-1}b$. $a(a^{-1}b) = b$. $(a^{-1})^{-1} = a$.

8 $e \cdot e = e$, so one such element exists. $a \cdot a = a \Rightarrow a^{-1}(a \cdot a) = a^{-1}a \Rightarrow a = e$.

9 $(ab)(b^{-1}a^{-1}) = [(ab)b^{-1}]a^{-1} = aa^{-1} = e$. $b^{-1}a^{-1}$ is the unique solution to $(ab)x = e$. If $a = (12)$ and $b = (13)$, then $(ab)^{-1} \neq a^{-1}b^{-1}$.

10 $(\mathbb{Z}, +), (\mathbb{Q}, +)$ and $(\mathbb{R}, +)$ are subgroups of $(\mathbb{C}, +)$.

11 The unique solution of $a^2 = a$ in H is in fact a solution in G. If e is the common identity and $a \in H$, then $ax = e$ has a unique solution in H and in G.

12 Check identity, inverses, closure.

13 $1 + 1, 1 + 1 + 1$, etc. and their inverses.

14 $2 + 2, 2 + 2 + 2$, etc. and their inverses.

15 $\alpha^2 = (13)(24)$, half-turn. $\beta\alpha = (13)$, reflection. $\langle\alpha\rangle = \{I, \alpha, \alpha^2, \alpha^3\}$.

16 3. $\langle\beta\rangle = \{I, \beta, \beta^2\}$.

17 order 1: (1); order 2: (12), (13), (23); order 3: (123), (132).

18 order 1: I; order 2: $\alpha^2, \beta, \beta\alpha, \beta\alpha^2, \beta\alpha^3$; order 4: α, α^3.

19 Obviously closed. g^0 is the identity. $g^{-n} = (g^n)^{-1}$. Yes.

20 $\alpha = (12345)$. $\langle\alpha\rangle$ has five elements.

21 Five elements. Order 1: I; order 5: a, a^2, a^3, a^4.

22 Six elements. Order 1: I; order 2: a^3; order 3: a^2, a^4; order 6: a, a^5.

23 $\{I\}, \{I, a^3\}, \{I, a^2, a^4\}, \{I, a, a^2, a^3, a^4, a^5\}$. All cyclic.

24 If $a = g^n$ and $b = g^m$, then $ab = g^{n+m} = g^{m+n} = ba$.

25 Since $a^i(a^{-1})^i = I$, $(a^{-1})^n = I$ and no lower power.

26 If $b \in H$, $b \neq ka$, then for some k, $ka < b < (k + 1)a$ and $0 < b - ka < a$, giving a smaller positive element of H.

27 If $G = \langle g \rangle$ then each subgroup contains an element g^a of least positive exponent. Then argue as in qn 26.

28 If $g^n = I$ then $g^r = g^{r+kn}$, so every element of $\langle g \rangle$ is in the given set. If $0 \leqslant r, s < n$ and $g^r = g^s$, then $g^{r-s} = g^{s-r} = I$ so either $r = s$ or the order of g is less than n. H contains $\langle g^a \rangle$ and so g^{la}. If H contains g^{la+k}, H contains g^k. Contradiction.

29 Given for $n = 1$. Assume $a^n b = ba^n$. Then
$$a^{n+1}b = (a^n \cdot a)b = a^n(ab) = a^n(ba) = (a^n b)a = (ba^n)a = ba^{n+1}.$$
E.g. $a^m ba^l = a^{m+l}b$, and $(ba^n)(ba^l) = a^{n+l}$. Closed. Identity a^0, inverse of ba^n is ba^{-n}.

33 $n = 1$, trivial. Assume $(ab)^n = a^n b^n$, then
$(ab)^{n+1} = (ab)^n(ab) = a^n b^n(ab) = a^n ab^n b$ by qn 29.

34 6. **35** 10. **36** 12. **37** mn/k.

38 $a = (12)(345)$, $b = (12)(67890)$.

42 $ba^2 b = baab = babbab = a^{-1} \cdot a^{-1}$. Likewise $ba^n b = a^{-n}$.

43 $G = \{a^i, ba^j \mid 0 \leqslant i, j < n\}$.

44 $(1) = (12)(12)$, $(123) = (12)(13)$, $(132) = (13)(12)$.

47 $(12)(13) = (123)$. $(134)(132) = (12)(34)$. In a string of transpositions every adjacent pair may be written either as a 3-cycle (if they overlap) or as a product of two 3-cycles (if they are disjoint).

48 A translation is the product of two reflections in parallel axes. The length of the translation is twice the distance between the axes. See qn 3.21.

49 See qn 3.37.

50

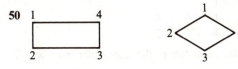

(1)	(1)	$z \mapsto z$	$(x, y) \mapsto (x, y)$
(12)(34)	(13)	$z \mapsto \bar{z}$	$(x, y) \mapsto (x, -y)$
(14)(23)	(24)	$z \mapsto -\bar{z}$	$(x, y) \mapsto (-x, y)$
(13)(24)	(13)(24)	$z \mapsto -z$	$(x, y) \mapsto (-x, -y)$.

51

$(123) \leftrightarrow (ace)(bdf)$ and $(23) \leftrightarrow (af)(be)(cd)$, for example.

52 $e \leftrightarrow I, a \leftrightarrow \alpha, a^2 \leftrightarrow \alpha^2$.

53 Two elements in the same row are ax and ay. $ax = ay \Rightarrow x = y$.

54 $\{I, p, q, r\}$.

55 $x \mapsto x \cdot 2$ is (123). $a \mapsto [x \mapsto x \cdot a]$ is an isomorphism.

57 If $\alpha : x \mapsto x \cdot a$, and $\beta : x \mapsto x \cdot b$, then $\alpha\beta : x \mapsto (x \cdot a) \cdot b = x \cdot (ab)$ which establishes the isomorphism.

62 $xy \mapsto \log xy = \log x + \log y$.

63 Each has order n.

64 $a^i \leftrightarrow b^i$.

65 If a, b are generators and $a^i \leftrightarrow b^i$, $a^m \cdot a^n = a^{m+n} \leftrightarrow b^{m+n} = b^m \cdot b^n$.

66 $a \cdot e = e \cdot a = a$, so $(a \cdot e)\alpha = (e \cdot a)\alpha = a\alpha$, so $a\alpha \cdot e\alpha = e\alpha \cdot a\alpha = a\alpha$ and $e\alpha$ is the identity.

67 $g \cdot g^{-1} = g^{-1} \cdot g = e$, so $(g \cdot g^{-1})\alpha = (g^{-1} \cdot g)\alpha = e\alpha$, so $g\alpha \cdot g^{-1}\alpha = g^{-1}\alpha \cdot g\alpha = e\alpha = $ identity.

68 $g^i\alpha = (g\alpha)^i$. $g^n = e \Rightarrow (g\alpha)^n = e\alpha$ so $g\alpha$ has order n or less. But $(g\alpha)^m = e\alpha \Rightarrow g^m\alpha = e\alpha$ and α is one–one so $g^m = e$. That is, elements of the same order must be matched under an isomorphism.

69

Number of elements of order	1	2	4
Cuboid	1	7	
Solid swastika	1	3	4
D_4	1	5	2

So no isomorphism possible, since elements of the same order must be matched by qn 68.

7

Inversions of the Möbius plane and stereographic projection

In this chapter we elaborate one example of a group generated by a set of elements, the group generated by all inversions, which includes the Möbius group as a subgroup, and we examine one particular isomorphism. The group generated by the inversions is in fact the full group of circle-preserving transformations of the plane, though we will not complete the proof of this since it depends on the fact that the identity function is the only function $\mathbb{R} \to \mathbb{R}$ which is both an isomorphism of the additive group of real numbers and also an isomorphism of the multiplicative group of real numbers. The isomorphism which we will study is between the full group of circle-preserving transformations of the Möbius plane, and a group of transformations of the surface of a sphere. The isomorphism is set up using stereographic projection and the interest in this isomorphism depends upon the preservation of geometric structure under this kind of projection.

Concurrent reading: Forder, pp. 19–29; Hilbert and Cohn-Vossen, pp. 248–68; Yaglom, pp. 55–7; Knopp chapters 3, 4 and 5; Hille, pp. 38–41, 46–58; Klein, chapter 2.

Inversion

1 If Σ is a circle with centre O and radius a and S is a circle which cuts Σ orthogonally and, moreover, a diameter of Σ meets S at the points A and B, what is the value of $OA \cdot OB$? (See qn 4.31.)

2 If Σ is a circle with centre O and radius a and A and B are two points on a radius of Σ with $OA \cdot OB = a^2$, then A and B are said to be *inverse points* with respect to Σ. Must any circle through A and B cut Σ orthogonally provided $A \neq B$?

3 If Σ is a circle and A and B are two points in the plane of Σ such that every circle through A and B meets Σ orthogonally, must A and B be inverse points with respect to Σ? If α is a Möbius transformation, what can you say about $A\alpha$ and $B\alpha$ in relation to $\Sigma\alpha$? What happens if $\Sigma\alpha$ is a straight line?

4 For any circle Σ of the plane, we define an *inversion* in Σ as a transformation of the Möbius plane which transposes every pair of inverse points and also transposes ∞ with the centre of Σ. In the case where Σ is the circle $|z| = 1$, does this definition coincide with that in qn 4.36? What are the fixed points of an inversion in Σ?

5 If S is a circle orthogonal to Σ, what is the image of S under the inversion in Σ?

6 If L is a diameter of Σ, what is the image of L under the inversion in Σ?

7 If z and z' are inverse points with respect to the circle $|z| = a$, use the working of qn 4.34 to show that if $z = re^{i\theta}$ then $z' = (a^2/r)\,e^{i\theta}$.

8 If two circles have the same centre and distinct radii, a and b, by taking the centre as the origin compute the product of successive inversions in the two circles. Give a geometric description of the composite transformation.

9 Prove that every direct similarity can be expressed as the product of two reflections and two inversions.

10 Use the fact that $z \mapsto 1/z$ is the product of a reflection and an inversion (as in qn 4.36) to prove that every Möbius transformation is a product of reflections and inversions, an even number in total.

11 Describe as many similarities or geometric analogies as you can between a reflection and an inversion in a circle. If we use the phrase 'inversion in a straight line' to mean a reflection, how does the group of transformations of the Möbius plane generated by the inversions relate to S_M and to the Möbius group?

12 If z and z' are inverse points with respect to a circle with centre s and radius R, show that if $z - s = re^{i\theta}$ then $z' - s = (R^2/r)e^{i\theta}$ and deduce that the inversion in this circle has the form $z \mapsto R^2/\overline{(z - s)} + s$. Show that for a suitable choice of a, b, c and d, this may be expressed in the form

$$z \mapsto \frac{a\bar{z} + b}{c\bar{z} + d},$$

with $ad - bc \neq 0$.

13 Show that the product of two inversions is a Möbius transformation.

14 What can be said about a transformation which is a product of an even number of inversions?

15 Give a general algebraic form for any transformation which can be expressed as a product of an odd number of inversions.

16 Must any transformation in the group generated by inversions be either a Möbius transformation or a product of a Möbius transformation and a reflection and so preserve the set of lines and circles in the Möbius plane? Any such transformation is called a circular transformation, and in fact the full group of circular transformations is generated by the inversions, though we will not prove this.

17 In the diagram of qn 4.1, prove that $NA \cdot NA' = NO^2$. Deduce the image of the circle on NO as diameter under an inversion in a circle with centre N and radius NO.

18 If A and B are points on the circumference of a circle Σ, what is the image of the line AB under the inversion in Σ?

19 If Σ is a circle with centre O and l is a line which does not cut Σ, let A be the foot of the perpendicular from O to l, and let B be any other point of l. Let A' be the inverse of A and B' be the inverse of B with respect to Σ. Show that B' lies on the circle on OA' as diameter. What is the inverse of l with respect to Σ?

Stereographic projection

Questions 20–28 establish that the image of a circle under stereographic projection is either a circle or a straight line. The method is to consider stereographic projection as part of a 3-dimensional inversion.

20 If Σ is a sphere with centre O and radius a, and we define inversion in Σ as a mapping of 3-dimensional real space with the point ∞ added such that two points on a radius of Σ, A and B are transposed when $OA \cdot OB = a^2$ and O is transposed with ∞, use qn 17 to determine the image of a tangent plane to Σ under an inversion in Σ.

21 Using the diagram of qn 4.1 and the definition of qn 4.35, use the notion of inversion in a sphere to describe stereographic projection.

22 With the help of qns 18 and 19 determine the image of a plane (other than a tangent plane) under inversion in a sphere.

23 What is the image of a sphere under inversion in a sphere?

24 In 3-dimensional real space, we let Σ denote the surface of a sphere with centre O and radius a, and we let P denote the foot of the perpendicular from O to a plane Π.
Determine the nature of $\Sigma \cap \Pi$

   (i) when $OP > a$,
   (ii) when $OP = a$,
   (iii) when $OP < a$.

25 Is every circle on a sphere the intersection of a plane with that sphere?

26 By considering stereographic projection as the restriction of a 3-dimensional inversion, determine the image of a circle under stereographic projection.

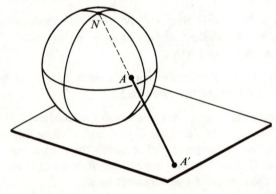

27 Using the notation of qn 4.35, determine the image of a circle through N under stereographic projection.

28 What is the image of the set of all circles on a sphere under stereographic projection?

Having established that under stereographic projection circles are projected onto circles or straight lines, we must establish one more fundamental property of stereographic projection, namely that it preserves angles of intersection. Access to a transparent or wire sphere will be helpful.

29 If two straight lines in the plane intersect in the point P', and P' is the stereographic projection of the point P, what can you say about the curves on the sphere which are projected onto the two given straight lines?

If two smooth curves intersect at a point P, then the angle between the curves is defined to be the angle between the tangents to the curves at P.

30 The first step in showing that stereographic projection preserves angles is to equate the angle between two lines in the image plane with the angle between two lines through the point of projection.

Let NS be the diameter of a sphere Σ and let Π_N and Π_S be the tangent planes to Σ at N and S respectively, so that Σ is like a ball between the two boards Π_N and Π_S. Are the planes Π_N and Π_S parallel?
Let l be a line in the plane Π_S, and let the plane containing the point N and the line l cut the plane Π_N in n. Why must the lines l and n be parallel? If l is the stereographic projection of the circle C on Σ, why must n be a tangent to C?

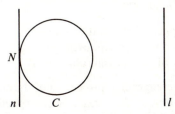

31 If l_1 and l_2 are lines in the plane Π_S and we define two corresponding lines n_1 and n_2 in the plane Π_N as in qn 30, why is the angle between l_1 and l_2 equal to the angle between n_1 and n_2? If the lines l_1 and l_2 are the stereographic projections of the circles C_1 and C_2, why is the angle between l_1 and l_2 equal to the angle of intersection of C_1 and C_2? What point on the sphere is projected to the point of intersection of l_1 and l_2?

From qn 31, the preservation of angles under stereographic projection is now immediate. The next step is to apply these two results, the preservation of circles and the preservation of angles under stereographic projection to the image of orthogonal circles under stereographic projection and hence to determine the transformation of the sphere which corresponds to the transformation of inversion in the plane.

32 If two circles on a sphere cut orthogonally, what can be said about their images under stereographic projection?

33 If a circle S and two points A and B on the surface of a sphere have images S', A' and B' respectively under stereographic projection, and A' and B' are inverse points with respect to S', what may be said about circles in the plane through A' and B' (see qn 2)? What may be deduced about circles on the sphere through A and B?

34 It is not easy to visualise orthogonal circles on the surface of a sphere until we have identified, for each circle on the sphere, a unique *tangent cone*. If P is a point on a circle S which lies on the surface of a sphere Σ, how many lines are there, through P, which are tangents to the sphere Σ and perpendicular to S at P?

35 If Σ is a sphere with centre O and radius a, and S is a circle lying on the surface of Σ with centre B and radius b, identify a point V on the line OB such that every tangent from V to Σ intersects S. The point V is called the *vertex* of the tangent cone to S.

36 If C and S are circles on the surface of the sphere Σ which cut orthogonally at the two points P and Q, what can be said about the tangents to the circle C at the points P and Q? What other point must lie in the plane of the circle C?

37 If every circle on the surface of a sphere through the two points A and B of the sphere meets the circle S on the sphere orthogonally, how must A and B relate to the vertex of the tangent cone to S?

38 If the circle S' in the plane is the stereographic projection of the circle S on the sphere, what transformation of the surface of the sphere would correspond to inversion in S' on the plane?

39 How must the answer to qn 38 be adapted if the circle S is a great circle, that is, an equator of the sphere?

40 Use the 3-dimensional analogue of qn 3.21 to say what you can about the transformation in the plane which corresponds to a rotation of the sphere about a diameter.

Riemann sphere

We now suppose that the sphere Σ has unit diameter, ON, and we let the tangent plane at O be a Gauss plane with 0 at O, and we label the points of $\Sigma - \{N\}$ with the set $\mathbb{C}$, using their images under stereographic projection from N as labels. We further label N with ∞, and refer to the sphere so labelled as the *Riemann sphere*.

41 Verify that a reflection of the Riemann sphere in the diametral plane perpendicular to ON is $z \mapsto 1/\bar{z}$.

42 Why must every rotation of the Riemann sphere be a Möbius transformation?

43 How are the points $z, \bar{z}, -z, -\bar{z}, 1/\bar{z}, 1/z, -1/\bar{z}, -1/z$ located on the Riemann sphere?

44 What is the point diametrically opposite z on the Riemann sphere?

45 If α is a rotation of the Riemann sphere, why must $(-1/\bar{z})\alpha = -1/\overline{(z\alpha)}$?

46 If
$$\alpha{:}z \mapsto \frac{az + b}{cz + d}$$
is a rotation of the Riemann sphere, by considering the images of 0 and ∞, prove that $\bar{a}b + \bar{c}d = 0$, and deduce that

$$c\bar{c}\left(\frac{bc - ad}{bc}\right) = a\bar{a} + c\bar{c}$$

and that

$$b\bar{b}\left(\frac{bc - ad}{bc}\right) = b\bar{b} + d\bar{d}.$$

By considering the images of 1 and -1, prove that $a\bar{a} + c\bar{c} = b\bar{b} + d\bar{d}$, and deduce that $b\bar{b} = c\bar{c}$ and $a\bar{a} = d\bar{d}$. Since $|a| = |d|$, we may put $d = \bar{a}e^{i\theta}$. Now deduce that $-b = \bar{c}e^{i\theta}$ so that

$$\alpha : z \mapsto \frac{az + b}{-\bar{b}e^{i\theta}z + \bar{a}e^{i\theta}}.$$

By multiplying top and bottom by a suitable number show that every rotation of the Riemann sphere has the form

$$z \mapsto \frac{az + b}{-\bar{b}z + \bar{a}}.$$

47 Let α denote the transformation of the Riemann sphere

$$z \mapsto \frac{az + b}{-\bar{b}z + \bar{a}}.$$

(i) Prove that the image of a pair of diametrically opposed points on the Riemann sphere under α is a pair of diametrically opposed points, that is, α preserves diameters.

(ii) Use qn 4.25 and qn 28 (the preservation of circles) to deduce that the image of a great circle under α is a great circle.

(iii) Use qn 4.44 to prove that α fixes a pair of diametrically opposed points, C and D, say.

(iv) Must the image of a circle through C and D, under α, be a circle through C and D?

(v) Identify a unique great circle orthogonal to all the circles through C and D. Use qns 4.42 and 31 to show that this circle is fixed by α.

(vi) What would be the image of the circle of (v) under a rotation of the sphere about the diameter CD?

(vii) Use the fact that the angle between two great circles through C and D is equal to the angle between their images under α to show that α coincides with a rotation at C, D and on the circle of (v).

(viii) Let P and Q be any two points on the Riemann sphere which are not diametrically opposed and let S be the centre of the sphere. If AB is the diameter perpendicular to the plane SPQ, by considering the preservation of great circles and angles under α, compare the spherical triangles APQ and $A\alpha P\alpha Q\alpha$, and deduce that the great circle arc PQ is equal in length to the great circle arc $P\alpha Q\alpha$ so that α acts as an isometry on the Riemann sphere.

48 If we let σ denote stereographic projection from the sphere to the plane, exhibit a function which establishes an isomorphism between the circular transformations of the Möbius plane and the circle-preserving transformations of the sphere.

Summary

Definition   Two points A and B are *inverse* with respect to a circle
qn 2   with centre O and radius a, if they lie on the same
         radius through O and $OA \cdot OB = a^2$. Also O and ∞ are
         inverse points.

Theorem   Two distinct points A and B are inverse with respect to
qn 3   a circle Σ if and only if every circle through A and B
         cuts Σ orthogonally.

Definition   An *inversion* of the Möbius plane with respect to a
qn 4   given circle transposes pairs of inverse points with res-
         pect to the given circle and also transposes the centre of
         the circle and ∞.

Theorem   Every Möbius transformation is a product of an even
qn 14   number of inversions.

Theorem   The group generated by the inversions maps the set of
qn 16   circles and lines onto the set of circles and lines.

Theorem   Under stereographic projection the set of circles on the
qn 28   sphere is mapped onto the set of circles and lines of the
         Möbius plane.

Theorem   The angle between two curves is preserved under stereo-
qn 31   graphic projection.

Theorem   A conical projection on the sphere corresponds to an
qn 38   inversion of the plane under stereographic projection.

Historical note

A. F. Möbius' study of the transformations generated by inversions (1852–56) included the complete enumeration of the circle-preserving transformations of the plane.

The connection between transformations of the plane and transformations of the surface of the sphere was identified by F. Klein in 1875 and fully explored in his *Lectures on the Icosahedron* (1884).

Answers to chapter 7

1 a^2.

2 Yes, see qn 4.32.

3 Möbius transformations preserve circles (qn 4.25) and orthogonality (qn 4.42) so $A\alpha$ and $B\alpha$ are inverse points with respect to $\Sigma\alpha$. If $\Sigma\alpha$ is a straight line, it is the perpendicular bisector of $A\alpha B\alpha$.

4 The points on the circumference of Σ.

5 S is mapped onto itself.

6 L is mapped onto itself with centre and ∞ interchanged.

8 $z \mapsto (b^2/a^2)z$, an enlargement.

9 If $a = r e^{i\theta}$, $z \mapsto az + b$ is the composite of $z \mapsto rz$ and $z \mapsto e^{i\theta}z + b$. $z \mapsto rz$ is the product of two inversions from qn 8. $z \mapsto e^{i\theta}z + b$ is the product of two reflections from qn 6.48.

10 Use qn 4.14.

11 Both have order 2. Both map circles and lines to circles and lines and preserve absolute values of angles of intersection. Inversions generate a subgroup of S_M. Even products of inversions give the Möbius group.

12 $a = s$, $b = R^2 - s\bar{s}$, $c = 1$, $d = -\bar{s}$.

13 Let
$$\alpha{:}z \mapsto \frac{a\bar{z} + b}{c\bar{z} + d} \quad \text{and} \quad \beta{:}\ z \mapsto \frac{A\bar{z} + B}{C\bar{z} + D}.$$
Calculate $\alpha\beta$.

14 It is in the Möbius group.

15 $z \mapsto \dfrac{a\bar{z} + b}{c\bar{z} + d}$ with $ad - bc \neq 0$.

16 Yes.

17 $NA = ON\cos\theta$, $NA' = ON\sec\theta$. $A \leftrightarrow A'$, so Σ is mapped to the tangent at O.

18 A circle through A, B and the centre of Σ.

19 Circle through $AA'B$ is orthogonal to Σ and so cuts OB at B'. Inverse of l is the circle on OA' as diameter.

20 Tangent plane at N is mapped to the sphere on ON as diameter.

21 Stereographic projection is the restriction of inversion in a sphere centre N, radius NO, to the sphere on NO as diameter.

22 A sphere through the centre of inversion.

23 Extend 16 to three dimensions. Image of a sphere is a sphere or a plane.

24 (i) When $OP > a$, $\Sigma \cap \Pi = \varnothing$. (ii) When $OP = a$, $\Sigma \cap \Pi = \{$one point$\}$.
(iii) When $OP < a$, $\Sigma \cap \Pi = $ a circle centre P.

25 Yes, because a circle is a plane figure.

26 A circle on Σ is the intersection of Σ and a plane Π. The image of Σ under inversion is the tangent plane; the image of Π from qn 22 is a sphere. Sphere and tangent plane intersect in a circle.

27 Plane section through N meets the tangent plane in a line.

28 The circles and lines of the plane.

29 They are both circles through N.

30 Yes, l is in Π_S, n is in Π_N but $\Pi_S \parallel \Pi_N$ so l and n do not meet. But they are coplanar, so they must be parallel. C and n are coplanar and n meets C at N. If n met C at an additional point, this point would project onto l and n would intersect l.

31 $l_1 \parallel n_1$, $l_2 \parallel n_2$. n_1 is a tangent to C_1, n_2 is a tangent to C_2. Angle between l_1 and $l_2 = $ angle between n_1 and $n_2 = $ angle between C_1 and C_2. The point of $C_1 \cap C_2$ distinct from N.

32 Images cut orthogonally.

33 Circles through A and B orthogonal to S.

34 Just one.

35

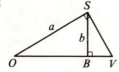

$$OV = a^2 / \sqrt{(a^2 - b^2)}$$

36 They lie in the plane of C and are perpendicular to S, so they pass through the vertex of the tangent cone to S.

37 Every plane through A and B cuts S orthogonally, and so contains the vertex of the tangent cone of S. So the vertex lies on AB.

38 Projection from the vertex of the tangent cone of S.

39 Tangent cone becomes tangent cylinder. Parallel projection. This amounts to reflection in the plane of the great circle.

40 Rotation of sphere = product of two reflections in diametral planes. Each of these reflections corresponds to an inversion in the plane. A rotation of the sphere corresponds to a product of two inversions.

42 Use qns 13 and 40.

43 At the vertices of a cuboid.

44 $-1/\bar{z}$.

45 Because diameters map to diameters.

48 $\alpha \mapsto \sigma\alpha\sigma^{-1}$.

8

Equivalence relations

This chapter puts in formal terms every child's instinctive activity of sorting a collection of things into different kinds. We will say that things count as being of the same kind when they are related by an equivalence relation, or they are equivalent. When we have gathered together all things of the same kind we call the set an equivalence class.

Concurrent reading: Green, chapter 2.

1 Two lines in a plane may be parallel, perpendicular or neither. If $l \parallel m$, does it follow that $m \parallel l$? If $l \perp m$, does it follow that $m \perp l$?
These results make both parallelism and perpendicularity *symmetric* relations.
Is orthogonality of circles a symmetric relation?
Is the divisibility of natural numbers a symmetric relation?

2 If $l \parallel m$ and $m \parallel n$, does it follow that $l \parallel n$? (We allow a line to be parallel to itself.) If $l \perp m$ and $m \perp n$, does it follow that $l \perp n$?
These results make parallelism, but not perpendicularity, a *transitive* relation.
Is orthogonality of circles a transitive relation?
Is divisibility of natural numbers a transitive relation?

3 If a set P is partitioned into disjoint subsets $A_1, A_2, A_3, \ldots$, where $A_i \cap A_j = \varnothing$ unless $i = j$, we call the A_i, *parts* of P and the set of all A_i a *partition* of P.
Check the three following results.
 (i) If $x \in P$, then x lies in the same part of P as x.
 (ii) If $x, y \in P$ and x lies in the same part of P as y, then y lies in the same part of P as x.
 (iii) If $x, y, z \in P$, and x lies in the same part of P as y, and y lies in the same part of P as z, then x lies in the same part of P as z.

4 If R is a relation on P which is

reflexive (i.e. $x \, R \, x$ for all $x \in P$),
symmetric (i.e. $x \, R \, y$ implies $y \, R \, x$) and
transitive (i.e. $x \, R \, y$ and $y \, R \, z$ imply $x \, R \, z$),

and for each $a \in P$ we define

$$R_a = \{x \,|\, x \, R \, a, \, x \in P\},$$

prove that
  (i) R_a is not empty,
  (ii) if $R_a \cap R_b$ is not empty, then $R_a \subseteq R_b$ and $R_b \subseteq R_a$, so $R_a = R_b$,
  (iii) the sets R_a for all $a \in P$ form a partition of P.

These sets are called the *equivalence classes* of P under the *equivalence relation R*.

5 We can exhibit a relation on a finite set by something that looks like a multiplication table, by listing the elements of the set across the top and down one side, and then the entries in the table ($\checkmark$ for yes, $\times$ for no) indicate whether the elements are related or not. Thus

	b
a	$\checkmark$

indicates that $a \, R \, b$.

For each of the following relations on the set $\{a, b, c\}$ determine whether it is reflexive, symmetric or transitive. If it is an equivalence relation, state the equivalence classes.

	a	b	c
a	$\times$	$\times$	$\times$
b	$\times$	$\checkmark$	$\checkmark$
c	$\times$	$\checkmark$	$\checkmark$

	a	b	c
a	$\checkmark$	$\checkmark$	$\checkmark$
b	$\checkmark$	$\times$	$\times$
c	$\checkmark$	$\times$	$\times$

	a	b	c
a	$\checkmark$	$\checkmark$	$\checkmark$
b	$\times$	$\checkmark$	$\checkmark$
c	$\times$	$\times$	$\checkmark$

	a	b	c
a	$\checkmark$	$\times$	$\times$
b	$\times$	$\checkmark$	$\times$
c	$\times$	$\times$	$\checkmark$

Exhibit two other tables for the same set, each of which defines an equivalence relation, and state the equivalence classes in each case.

6 The most familiar equivalence relation is that of equality ($=$) on a set. What are the equivalence classes in this case?

7 A relation R is defined on the set of integers $\mathbb{Z}$ by $a \, R \, b$ when $a - b$ is an even number.
  (i) Check that R is reflexive, symmetric and transitive (that is, R is an equivalence relation).
  (ii) Find the sets R_1, R_2, R_3 and R_4 according to the notation of qn 4. The two classes here are called residue classes modulo 2 and $a \, R \, b$ is usually written $a \equiv b \pmod 2$ in this case.

8 A relation R is defined on the set of complex numbers, $\mathbb{C}$, by $z \, R \, w$ when $|z| = |w|$.

(i) Check that R is an equivalence relation.

(ii) Illustrate the sets R_1 and R_2 in a diagram of the Gauss plane.

9 In the group of isometries of the plane, we denote the subgroup fixing 0 (consisting of the rotations about 0 and the reflections with axes through 0) by M_0, and we use this group to define a relation R on the points of the plane by

$A \, R \, B$ when $A\mu = B$ for some $\mu \in M_0$.

(i) Prove that R is an equivalence relation.

(ii) Find the equivalence class containing a point distant 1 unit from 0.

(iii) Identify the equivalence classes, which, in this case, are called *orbits* of the group M_0.

10 If G is the group of symmetries of a regular pentagon, draw a regular pentagon and mark the possible images of a vertex under the ten elements of the group. These form the orbit of a vertex.

Mark another point P of the pentagon and the images of this point under the ten elements of G. These points form the orbit of P under G.

11 If G is any group of permutations on a set A, and we define a relation R on A by

$a \, R \, b$ when $a\alpha = b$ for some $\alpha \in G$,

prove that R is an equivalence relation on A.

The equivalence classes of A under R are called the *orbits* of G. When G has only one orbit, G is said to be transitive (qn 3.54).

12 If $\alpha \in S_n$, what is the ordinary name for the orbits of $\langle \alpha \rangle$?

13 If G is the group of enlargements of the plane with centre O, illustrate the orbits of G in a diagram.

14 If G is the group of translations of the Euclidean plane parallel to the real axis, illustrate the orbits of G in a diagram.

15 On the framework of a regular tetrahedron mark a point which is neither a vertex nor the midpoint of an edge, and identify all the points of the same orbit under the group of rotational symmetries of the tetrahedron.

The *length* of an orbit is the number of points it contains.

What is the length of the orbit to which a vertex belongs? What is the length of the orbit to which the midpoint of an edge belongs?

16 Sketch a framework of two regular tetrahedra sharing one face. Identify the orbits to which the vertices belong under the group of rotational symmetries of the figure.

Summary

Definition A relation R on a set A is said to be *reflexive* if $a R a$
qn 4   for all $a \in A$.

A relation R on a set A is said to be *symmetric* if
$a R b \Rightarrow b R a$.

A relation R on a set A is said to be *transitive* if $a R b$
and $b R c \Rightarrow a R c$.

A relation on a set A which is reflexive, symmetric and transitive is said to be an *equivalence* relation.

Theorem If R is an equivalence relation on a set A, then those
qn 4   subsets of A whose elements are related under R are called *equivalence classes* under R and these partition A.

Theorem If G is a permutation group on a set S, and a relation R
qn 11   is defined on S by $a R b$ when there exists an $\alpha \in G$ such that $a\alpha = b$, then R is an equivalence relation on S, and the equivalence classes are called the *orbits* of G.

Historical note

J. L. Lagrange (1773) when classifying quadratic forms in the theory of numbers used the term 'equivalent' to describe those in the same class. It is however the use of the symbol $\equiv$ for congruence of integers by C. F. Gauss in his *Disquisitiones Arithmeticae* (1801) which is usually taken as the birth of the modern notion of an equivalence relation. The modern symbolism and style of description come from the work of A. N. Whitehead and B. Russell (1910).

Answers to chapter 8

1 Orthogonality symmetric, divisibility not.

2 Divisibility transitive, orthogonality not.

4 (i) $a R a$, reflexive. (ii) Let $c \in R_a \cap R_b$, so $c R a$ and $c R b$. $c R a \Rightarrow a R c$, symmetric. $x \in R_a \Rightarrow x R a$, with $a R c \Rightarrow x R c$, transitive; with $c R b \Rightarrow x R b$, transitive; so $x \in R_b$ and $R_a \subseteq R_b$. (iii) Each a in R_a. From (ii) the R_a are identical or disjoint.

5 Only the last is an equivalence relation. Make further equivalence relations either by adding $a R a$ to the first table, or by having $x R y$ for all x and all y.

6 Singleton sets.

7 (ii) $R_1 = R_3 = \{\text{odd numbers}\}$. $R_2 = R_4 = \{\text{even numbers}\}$.

8 Circles centre 0, radius 1, 2 respectively.

9 (i) $AI = A$, reflexive. $A\mu = B \Rightarrow B\mu^{-1} = A$, symmetric. $A\mu = B$ and $B\nu = C$ implies $A\mu\nu = C$, transitive. (ii) Circle centre 0, radius 1. (iii) Circles centre 0.

11 Argue as in qn 9(i).

12 Cycles.

13 Lines through O excluding O.

14 Lines parallel to the real axis.

15 Orbit of general point has length 12; orbit of vertex has length 4; orbit of midpoint has length 6.

16 One orbit of length 2, one orbit of length 3.

9

Cosets

When a subgroup of a group G is given, the subgroup provides a way of partitioning the group into disjoint sets, all of the same size. This partition is called a partition into (left or right) cosets of the subgroup. The fact that the cosets are all of the same size enables us to show that the number of elements in a subgroup must be a factor of the number of elements in the whole group when the group contains a finite number of elements in all.

Concurrent reading: Birkhoff and MacLane, chapter 6, section on Lagrange's theorem; Green, chapter 6, sections 1–5; Fraleigh, section 11.

1 A relation R is defined on the group $(\mathbb{Z}, +)$ by $a\,R\,b$ when $a - b \in \langle 3 \rangle$. Prove that R is an equivalence relation and find the equivalence classes. The three classes are called the *residue classes modulo* 3 and $a\,R\,b$ in this case is usually written $a \equiv b \pmod 3$.

2 A relation R is defined on the group $(\mathbb{Z}, +)$ by $a\,R\,b$ when $a - b \in \langle n \rangle$. Prove that R is an equivalence relation and find the equivalence classes. The n classes are called the *residue classes modulo n* and $a\,R\,b$ in this case is usually written $a \equiv b \pmod n$.

3 If the group C_6 is generated by the element a and a relation R is defined on C_6 by

$x\,R\,y$ when $xy^{-1} \in \langle a^3 \rangle$,

prove that R is an equivalence relation and find the equivalence classes.

4 If H is a subgroup of a group G and a relation R is defined on G by

$x\,R\,y$ when $xy^{-1} \in H$, prove that R is an equivalence relation and that H is one of the equivalence classes.

The equivalence classes here are called *cosets* or, more precisely, the *right cosets* of H in G.

5 With the notation of qn 4, show that if $x\,R\,a$, then $x = ha$ for some $h \in H$.

6 With the notation of qns 4 and 8.4, show that if $H = \{h_1, h_2, h_3, \ldots, h_n\}$, then R_a, the right coset of H containing a is $\{h_1a, h_2a, h_3a, \ldots, h_na\}$, and that R_a contains exactly the same number of elements as H.
   Because of this result, the right coset of H containing a is usually denoted by Ha.

7 If D is the group of direct isometries of the Euclidean plane and T is the translation group, describe the elements lying in a right coset of T in D.

8 If Δ is the group of dilatations of the Euclidean plane (qn 3.59) and T is the translation group, describe the elements lying in a right coset of T in Δ.

9 If we denote the number of distinct elements in a finite set S by $|S|$, with the notation of qn 6, must $|H| = |Ha|$? Must any two right cosets of a finite subgroup contain the same number of elements?

h_1	h_1a	h_1b	
h_2	h_2a	h_2b	
h_3	h_3a	h_3b	
$\vdots$	$\vdots$	$\vdots$	
h_n	h_na	h_nb	

10 If G is a finite group with a subgroup H, use the fact that the right cosets of H partition G (qn 8.4) to show that $|H|$ is a factor of $|G|$. (*Lagrange's theorem*)
   The number $|G|$ is called the *order* of the group G and $|H|$ the *order* of the subgroup H.

11 If a group contains an element of order n (qn 6.16), what is the order (qn 10) of the cyclic subgroup which it generates?

12 If G is a group of order 12, what possible orders of subgroups could G have? What possible orders may the elements of G have? By considering the group A_4, verify that not every possibility need occur. (Just concentrate on the orders of elements now, we will look at the possible subgroups later.)

13 If H is a subgroup of G, and the number of right cosets of H in G is finite, the number of right cosets of H is called the *index* of H in G and denoted by $|G:H|$. Find the index of H in G in the following cases.

(i) $G = \langle a \rangle = C_6$ and $H = \langle a^2 \rangle$.

(ii) $G = \langle a \rangle = C_6$ and $H = \langle a^3 \rangle$.

(iii) $G = S_3$ and $H = A_3$.

(iv) $G = S_3$ and $H = \{(1), (23)\}$.

(v) $G = D_4 = \langle a, b \rangle$, where $a^4 = b^2 = e$ and $bab = a^{-1}$, and $H = \langle a \rangle$.

14 After labelling the vertices of a regular pentagon with the numbers 1, 2, 3, 4 and 5, express each of the symmetries of the pentagon as a permutation of these vertices. Subdivide the elements of this symmetry group into those elements which map $1 \mapsto 1$, those which map $1 \mapsto 2$, those which map $1 \mapsto 3$, those which map $1 \mapsto 4$ and those which map $1 \mapsto 5$. Use qn 2.40 to name the subset mapping $1 \mapsto 1$. Must it be a subgroup? If α and β are symmetries of the regular pentagon with $1\alpha = 2$ and $1\beta = 2$, find $1\alpha\beta^{-1}$ and hence describe the subset of elements mapping $1 \mapsto 2$. Name also the subset of elements mapping $1 \mapsto 3$. How does the orbit of 1 under this group relate to the subgroup stabilising 1?

15 After labelling the vertices of a regular tetrahedron with the numbers 1, 2, 3 and 4, express each of the rotational symmetries of the tetrahedron as a permutation of these vertices. Subdivide the elements of this symmetry group into those elements which map $1 \mapsto 1$, those which map $1 \mapsto 2$, those which map $1 \mapsto 3$ and those which map $1 \mapsto 4$. Does this subdivision partition the group? Do the elements fixing 1 form a subgroup? If α and β are rotational symmetries of the tetrahedron such that $1\alpha = 2 = 1\beta$, find $1\alpha\beta^{-1}$ and name the subset mapping $1 \mapsto 2$.

16 Write down the subgroup stabilising 1 in S_4. Call this subgroup H. Calculate the six elements of the coset $H(12)$, and find the image of 1 under each of the elements in this coset.

17 If G is a group of permutations of a set A (containing the element 1) and we define a mapping $G \to A$ by $\alpha \mapsto 1\alpha$ for each $\alpha \in G$,

(i) describe the image set of this mapping and give the special name of this image set;

(ii) describe the full subset of G which has the image 1 and give the special name of this subset;

(iii) if two elements α, β of G have the same image under this function, what can be said about the element $\alpha\beta^{-1}$?

If G_1 denotes the subgroup of G which stabilises 1, prove that the function given by

$G_1\alpha \mapsto 1\alpha$

with domain the right cosets of G_1 and range the orbit of 1 is well defined and one–one.

18 Use Lagrange's theorem to describe the length of the orbit of 1 in a permutation group on a set of elements including 1.

19 If $\alpha = (1234)(567)(89)$ and $C_{12} = \langle\alpha\rangle = G$, identify the orbits of G, and the stabilisers of each of the digits 1, 2, 3, 4, 5, 6, 7, 8, 9 in G and illustrate the result of qn 18.

20 Illustrate qn 18 by considering the group of rotational symmetries of a cube, and identifying the stabiliser and orbit of (i) a vertex, (ii) a face, (iii) an edge.

21 Find the stabiliser of 0 in the group D of direct isometries of the Euclidean plane. Use qn 17 to describe the elements in a right coset of this stabiliser.

22 If H is a subgroup of any group G, prove that the relation R defined by $x\,R\,y$ when $x^{-1}y \in H$ is an equivalence relation on G.
The equivalence classes under this relation are called the *left cosets* of H in G.

23 With the notation of qn 22, show that if $x\,R\,a$, then $x = ah$ for some $h \in H$. Conversely, for any $h \in H$, prove that $ah\,R\,a$.
If $H = \{h_1, h_2, h_3, \ldots, h_n\}$ exhibit the equivalence class containing a. This equivalence class is usually denoted by aH.

24 If a and b are elements of a group G containing a finite subgroup H, must $|aH| = |H| = |Ha| = |Hb| = |bH|$?

25 Could we have proved qn 10 using left cosets instead of right cosets?

26 Is the index of H in G equal to the number of *left* cosets of H?

27 In the group $D_4 = \langle a, b\rangle$, where $a^4 = b^2 = e$ and $bab = a^{-1}$, exhibit all the left cosets and all the right cosets of each of the subgroups $\{e, a^2\}$ and $\{e, b\}$.
When in a group G with a subgroup H, every left coset of H is also a right coset of H, then the subgroup H is said to be *normal*. Are either or both of the subgroups $\{e, a^2\}$ and $\{e, b\}$ normal?

28 If H is any subgroup of a group G, prove that $a \in Ha$ and $a \in aH$ for every $a \in G$.
If H is a normal subgroup of a group G, prove that $aH = Ha$ for every $a \in G$.
If H is a subgroup but not a normal subgroup of a group G, prove

that for some $a \in G$ $aH \neq Ha$. Check your definition of a normal subgroup.

29 If D is the group of direct isometries of the Euclidean plane, T is the translation group and R is the group of rotations with centre 0, find out whether either T or R are normal subgroups of D. Just one α would be enough to show $R\alpha \neq \alpha R$, but the contrary would have to hold for all direct isometries α.

30 If Δ is the group of dilatations of the Euclidean plane, T is the translation group and E is the group of enlargements with centre 0, find out whether T or E are normal subgroups of Δ.

31 Fill in the Cayley table for the group D_4 (of qn 27) as shown here, with the cosets of $\{e, a^2\}$ grouped together.

	e a^2	a a^3	b ba^2	ba ba^3
e				
a^2				
a				
a^3				
b				
ba^2				
ba				
ba^3				

32 If H and K are subsets of a group G, we define $HK = \{hk | h \in H, k \in K\}$.

If H is a normal subgroup of a group G prove that every element in $H(ab)$ lies in $(Ha)(Hb)$, and also that every element in $(Ha)(Hb)$ lies in $H(ab)$, so that $Ha \cdot Hb = Hab$.

Summary

Definition If H is a subgroup of a group G, then the relation R
qn 4 defined on G by $x R y$ when $xy^{-1} \in H$ is an equivalence relation, and the equivalence classes are called the *right cosets* of H.

Theorem Every right coset of a subgroup H has the form
qn 6 $Ha = \{ha | h \in H\}$. If G is a finite group, each right
qn 10 coset of H contains the same number of elements as H, and so $|G| = |H| \cdot$ (the number of right cosets of H). *Lagrange's theorem.*

Definition If H is a subgroup of G and H has a finite number of
qns 13, 26 cosets in G, then the number of right cosets equals the

number of left cosets and this number is called the *index* of H in G.

Theorem   If G is a permutation group on a set S, $1 \in S$, and G_1 is
qn 17   the stabiliser of 1 in G, there is a one–one correspondence between the right cosets of G_1 and the points
in the orbit of 1.

Definition   If H is a subgroup of a group G, then the relation R
qn 22   defined on G by $x\,R\,y$ when $x^{-1}y \in H$ is an equivalence
relation, and the equivalence classes are called the *left
cosets* of H.

Definition   If N is a subgroup of a group G for which every left
qn 27   coset is a right coset, then N is called a *normal* subgroup
of G.

Definition   If A and B are subsets of a group G, then
qn 32 $AB = \{ab | a \in A, b \in B\}$.

Theorem   If N is a normal subgroup of a group G then
qn 32 $(Na)(Nb) = N(ab)$.

Historical note

In 1770, J. L. Lagrange affirmed that the number of distinct values
which a function of n variables $f(x_1, x_2, \ldots, x_n)$ could assume if the
variables were given constant values and then permuted was a factor of
$n!$. This was the nearest he got to a statement of 'Lagrange's theorem'.
The point is that the subset of permutations of the suffices which keeps
the value of the function constant forms a subgroup of S_n, say H, and
the subset changing the value to some other particular value forms a
right coset of H. Thus, the different values of the function are in
one–one correspondence with the cosets of H. The result which
Lagrange had claimed was proved by P. Abbati in 1802 who also knew
that the converse was false. Abbati's proof involved a rectangular array
of cosets, something which first appeared in the work of L. Euler
(1760) on multiplicative groups in modular arithmetic. The statement
and proof of Lagrange's theorem for any finite group of permutations
appears in the third edition of J. A. Serret's textbook (1866). It was the
identification of normal subgroups and their significance by E. Galois
(1830) which stimulated much of the later work on groups. The term
'coset' was coined by G. A. Miller in 1910.

Answers to chapter 9

1 $3n\,R0$, $3n + 1\,R1$, $3n + 2\,R2$.

2 $0 \in \langle n \rangle$, so $a\,Ra$. $a - b \in \langle n \rangle \Rightarrow b - a \in \langle n \rangle$, so $a\,Rb \Rightarrow b\,Ra$. $a - b$ and $b - c \in \langle n \rangle \Rightarrow (a - b) + (b - c) \in \langle n \rangle$, so $a\,Rb$ and $b\,Rc \Rightarrow a\,Rc$.

3 $\{e, a^3\}$, $\{a, a^4\}$, $\{a^2, a^5\}$.

4 $e \in H \Rightarrow xx^{-1} \in H \Rightarrow x\,Rx$. $xy^{-1} \in H \Rightarrow (xy^{-1})^{-1} \in H \Rightarrow yx^{-1} \in H$, so $x\,Ry \Rightarrow y\,Rx$. xy^{-1} and $yz^{-1} \in H \Rightarrow (xy^{-1})(yz^{-1}) \in H \Rightarrow xz^{-1} \in H$, so $x\,Ry$ and $y\,Rz \Rightarrow x\,Rz$. $H = R_e$.

5 $x\,Ra \Rightarrow xa^{-1} \in H \Rightarrow xa^{-1} = h \Rightarrow x = ha$.

6 From qn 5, $x \in R_a \Rightarrow x = ha$. Also $x = ha \Rightarrow xa^{-1} \in H \Rightarrow x\,Ra \Rightarrow x \in R_a$. So $R_a = \{h_1 a, h_2 a, \ldots, h_n a\}$. The mapping $H \to R_a$ given by $h \mapsto ha$ is a bijection.

7 Using the notation of qn 6, take $T = H$ and take $a : z \mapsto -z$. Then every element in the coset Ha has the form $z \mapsto -z + c$. The coset consists of all the half-turns. Generally, all the rotations through the same angle lie in the same coset of T.

8 All enlargements by the same scale factor form a coset.

11 An element of order n generates a cyclic subgroup of order n.

12 Subgroups may have order 1, 2, 3, 4, 6 or 12. Elements may have just these orders. All elements of A_4 have order 1, 2 or 3.

13 (i) 2, (ii) 3, (iii) 2, (iv) 3, (v) 2.

14 (1), (25)(34); stabiliser of 1;
(12345), (12)(35);
(13524), (13)(45);
(14253), (14)(23);
(15432), (15)(24).
$\alpha\beta^{-1}$ lies in stabiliser of 1, so α and β lie in the same right coset. Each point of the orbit matches one coset.

15 A right coset of the stabiliser of 1.

16 $H(12) = \{(12), (1234), (1243), (123), (124), (12)(34)\}$.

17 (i) The orbit of 1. (ii) The stabiliser of 1. (iii) $\alpha\beta^{-1}$ lies in the stabiliser of 1. $G_1\alpha = G_1\beta \Leftrightarrow 1\alpha = 1\beta$.

18 Length of orbit $=$ index of stabiliser.

19 Orbit $\{1, 2, 3, 4\}$: stabiliser $\{(1), (567), (576)\}$.
Orbit $\{5, 6, 7\}$: stabiliser $\{(1), (1432)(89), (13)(24), (1234)(89)\}$.

Orbit $\{8, 9\}$: stabiliser $\{(1), (13)(24)(576), (567), (13)(24), (576), (13)(24)(567)\}$.

20 (i) Length of orbit 8, order of stabiliser 3. (ii) Length of orbit 6, order of stabiliser 4. (iii) Length of orbit 12, order of stabiliser 2.

21 $z \mapsto e^{i\theta}z$. All rotations and translations carrying $0 \mapsto c$.

22 Argument similar to qn 4.

23 $aH = \{ah_1, ah_2, ah_3, \ldots, ah_n\}$.

25, 26 Yes.

27 Left and right cosets of $\{e, a^2\}$; $\{a, a^3\}$, $\{b, ba^2\}$, $\{ba, ba^3\}$.
Left cosets of $\{e, b\}$; $\{a, ab = ba^3\}$, $\{a^2, a^2b = ba^2\}$, $\{a^3, a^3b = ba\}$.
Right cosets of $\{e, b\}$; $\{a, ba = a^3b\}$, $\{a^2, a^2b = ba^2\}$, $\{a^3, ba^3 = ab\}$.

28 $a = e \cdot a = a \cdot e$. If H is normal, the left coset aH is equal to one right coset, but right cosets are identical or disjoint so $aH = Ha$. If $aH = Ha$ for all a, then, plainly, H is a normal subgroup.

29 T is normal, R is not.

30 T is normal, E is not.

32 $x \in H(ab) \Rightarrow x = h \cdot a \cdot b = ha \cdot eb \in (Ha)(Hb)$ whether H is normal or not. $x \in (Ha)(Hb) \Rightarrow x = h_1ah_2b$, but H normal implies $ah_2 = h_3a$ and so $x = (h_1h_3)ab \in Hab$.

10

Direct product

This chapter introduces a simple method of constructing a new group from two given groups, which is useful both for construction and for analysis. The principal use that we will make of this construction, in later chapters, is the construction of groups of vectors. Qn 11 gives the key theorem for decomposition.

Concurrent reading: Fraleigh, section 8.

1 If $(A, \cdot)$ is the cyclic group $C_2 = \{e, a\}$, the cartesian product $A \times A$ consists of the four elements (e, e), (e, a), (a, e) and (a, a). Make a table to exhibit all possible products of these elements under the operation defined by
$$(b_1, c_1)(b_2, c_2) = (b_1 \cdot b_2, c_1 \cdot c_2).$$
Is the table that of a group isomorphic to D_2? The group constructed in this way is called the *direct product* $C_2 \times C_2$.

2 By analogy with qn 1, construct the direct product $A \times B$, where A is the cyclic group $C_2 = \{e, a\}$ and B is the cyclic group $C_3 = \{e, b, b^2\}$. Find the order of the element (a, b) in $C_2 \times C_3$ and determine whether $C_2 \times C_3$ is cyclic.

3 As in qn 2, construct the direct product $C_2 \times C_5$ and show that this is a cyclic group.

4 Construct the direct product $C_2 \times C_4$ and find the order of each element in this group. Is it a cyclic group?

5 Give a formal definition of the direct product of two groups $(A, \cdot)$ and $(B, \circ)$ which generalises qns 1–4. Verify that your definition must give a group.

6 If A is the multiplicative group of real numbers $(\mathbb{R} - \{0\}, \cdot)$ and B

is the additive group of real numbers $(\mathbb{R}, +)$ what is $(a, b)(c, d)$ in the direct product $A \times B$?

7 With the notation of qn 6 does the operation defined by

$(a, b)(c, d) = (ac, bc + d)$

make $A \times B$ a group? a direct product?

8 Find a subgroup of the direct product $A \times B$ which is isomorphic to A and a subgroup isomorphic to B.

9 If $A = \{a_1, a_2, a_3, a_4\}$ and $B = \{b_1, b_2, b_3, b_4, b_5, b_6\}$ and a_1 is the identity in the group $(A, \cdot)$ and b_1 is the identity in the group $(B, \cdot)$, use chapter 9 to describe the rows and columns of the array below.

(a_4, b_1) (a_4, b_2) (a_4, b_3) (a_4, b_4) (a_4, b_5) (a_4, b_6)
(a_3, b_1) (a_3, b_2) (a_3, b_3) (a_3, b_4) (a_3, b_5) (a_3, b_6)
(a_2, b_1) (a_2, b_2) (a_2, b_3) (a_2, b_4) (a_2, b_5) (a_2, b_6)
(a_1, b_1) (a_1, b_2) (a_1, b_3) (a_1, b_4) (a_1, b_5) (a_1, b_6)

10 Show that the group $G = \langle(1234), (56)\rangle$ is isomorphic to the direct product of $A = \langle(1234)\rangle$ and $B = \langle(56)\rangle$.

11 If a group G contains subgroups A and B then assuming
  (i) each $g \in G$ may be expressed uniquely in the form $g = ab$, where $a \in A$ and $b \in B$, and
  (ii) $ab = ba$ for $a \in A$ and $b \in B$,
prove that G is isomorphic to the direct product $A \times B$.

12 Exhibit the necessity for condition (ii) in qn 11 by considering $G = S_3$, $A = \langle(123)\rangle$ and $B = \langle(23)\rangle$, showing that the first condition (i) of qn 11 holds, but that condition (ii) fails and G is not isomorphic to $A \times B$, from qn 2.

13 If A, B and C are groups, define the direct product $A \times B \times C$ as $(A \times B) \times C$ and exhibit the product of two typical elements.

14 If the six faces of a cuboid are labelled 1, 2, 3, 4, 5 and 6 with $\{1, 2\}$, $\{3, 4\}$ and $\{5, 6\}$ labelling pairs of opposite faces, list the eight symmetries of the cuboid as permutations of its faces and show that this group of symmetries is isomorphic to the direct product of the groups $\langle(12)\rangle$, $\langle(34)\rangle$ and $\langle(56)\rangle$.

15 A bijection of $\mathbb{R} \times \mathbb{R}$ onto the complex numbers $\mathbb{C}$ is given by $(x, y) \mapsto x + iy$. If this is an ismorphism of a group onto the group $(\mathbb{C}, +)$, how would you describe the domain? In this case the domain with its implied operation of vector addition is known as the *direct sum* $\mathbb{R} \oplus \mathbb{R}$.

Summary

Definition If $(A, \cdot)$ and $(B, \circ)$ are groups then the group formed by
qn 5   the cartesian product $A \times B$ under the operation defined by

$$(a_1, b_1)(a_2, b_2) = (a_1 \cdot a_2, b_1 \circ b_2)$$

is called the *direct product* of the groups A and B.

Theorem The direct product $A \times B$ contains a normal subgroup
qn 9   isomorphic to A and a normal subgroup isomorphic to
B.

Theorem A group G is isomorphic to the direct product of two
qn 11 subgroups A and B when each element of G may be
uniquely expressed as a product of an element in A and
an element in B, and $ab = ba$ for $a \in A$ and $b \in B$.

Historical note

Appolonius (200 B.C.), in his study of the conic sections, used two numbers to denote the position of a point in a plane. This idea was exploited by R. Descartes (1639) in his study of plane curves, but even for Descartes, coordinates were always positive numbers. The notation of ordered pairs and ordered triples to describe the position of points in space was developed during the eighteenth century. The formal definition of a cartesian product is due to G. Cantor at the end of the nineteenth century.

The earliest study of groups formed as a direct product was A. L. Cauchy's study of intransitive groups of permutations (1845). Groups of vectors are implicit in A. Cayley's work on quantics and appear explicitly in the form of translations in C. Jordan's *Traité des Substitutions* (1870).

Answers to chapter 10

1 (e, e) (e, a) (a, e) (a, a)
(e, a) (e, e) (a, a) (a, e)
(a, e) (a, a) (e, e) (e, a)
(a, a) (a, e) (e, a) (e, e)

2 (a, b) has order 6, so $C_2 \times C_3$ is a cyclic group.

3 If $C_2 = \{e, a\}$ and $C_5 = \{e, b, b^2, b^3, b^4\}$ then (a, b) has order 10.

4 Three elements of order 2 and four of order 4. Not cyclic.

5 Direct product of $(A, \cdot)$ and $(B, \circ)$ is the cartesian product $A \times B$ with operation defined by $(a_1, b_1)(a_2, b_2) = (a_1 \cdot a_2, b_1 \circ b_2)$. Closure implicit in definition. Associativity immediate. Identity is (e, e). Inverse of (a, b) is (a^{-1}, b^{-1}).

6 $(ac, b + d)$.

7 Not a direct product, but still a group with identity $(1, 0)$. Inverse of (a, b) is $(a^{-1}, - ba^{-1})$. Compare qn 3.3.

8 $A \times \{e\}$ is isomorphic to A. $\{e\} \times B$ is isomorphic to B.

9 Rows are cosets (left and right) of $\{e\} \times B$. Columns are cosets of $A \times \{e\}$.

10 Match ab with (a, b). Use qn 6.31.

11 $(a_1, b_1) \mapsto a_1 b_1$, $(a_2, b_2) \mapsto a_2 b_2$. From (i) $a_1 b_1 = a_2 b_2 \Rightarrow a_1 = a_2$ and $b_1 = b_2$ so mapping is one–one. $(a_1 a_2, b_1 b_2) \mapsto a_1 a_2 b_1 b_2 = a_1 b_1 a_2 b_2$ by (ii) so mapping is structure-preserving.

12 $A \times B = C_3 \times C_2$ which is isomorphic to C_6 from qn 2.

13 $(a_1, b_1, c_1)(a_2, b_2, c_2) = (a_1 a_2, b_1 b_2, c_1 c_2)$.

14 Half-turns $(34)(56)$, $(12)(56)$, $(12),(34)$. Reflections (12), (34), (56). Point reflection $(12)(34)(56)$. Extend idea of qn 11 to qn 13.

15 The domain is the direct product of $(\mathbb{R}, +)$ with itself.

11

Fields and vector spaces

The word 'field' describes abstractly an algebraic structure in which the operations $+$, $-$, $\times$ and $\div$ may be performed in much the same way as they are performed with the rational numbers. When copies of a field F are used to make cartesian products such as $F \times F$ or $F \times F \times F$ such cartesian products are called vector spaces when furnished with a suitable addition of elements and multiplication by a scalar (i.e. a field element).

Concurrent reading: Birkhoff and MacLane, chapter 2, section 1 and chapter 7, sections 1–4.

Fields

If in a group $(G, \cdot)$ $xy = yx$ for any two elements $x, y \in G$, then the group G is said to be commutative, or *abelian*.

1 The rational numbers $\mathbb{Q}$, the real numbers $\mathbb{R}$ and the complex numbers $\mathbb{C}$ each form an abelian group under addition, and, with the deletion of zero, an abelian group under multiplication. These two properties, together with the distributive laws $a \cdot (b + c) = a \cdot b + a \cdot c$ and $(a + b) \cdot c = a \cdot c + b \cdot c$ make $(\mathbb{Q}, +, \cdot)$, $(\mathbb{R}, +, \cdot)$ and $(\mathbb{C}, +, \cdot)$ into *fields*. why do the integers $(\mathbb{Z}, +, \cdot)$ *not* form a field?

2 Does the set $\{0, 1\}$ with the operations $+$ and $\cdot$ given by

+	0	1		·	0	1
0	0	1		0	0	0
1	1	0		1	0	1

form a field? This set with the operations given is called $\mathbb{Z}_2$.

3 Does the set $\{0, 1, 2\}$ with the operations $+$ and $\cdot$ given by

+	0	1	2
0	0	1	2
1	1	2	0
2	2	0	1

$\cdot$	0	1	2
0	0	0	0
1	0	1	2
2	0	2	1

form a field? This set, with the given operations is called $\mathbb{Z}_3$. The distributive law $a(b + c) = ab + ac$ for $a \neq 0$ is equivalent to the claim that the mapping $x \mapsto ax$ is an isomorphism of $(\mathbb{Z}_3, +)$ onto $(\mathbb{Z}_3, +)$.

4 If $(F, +, \cdot)$ is any field and 0 is the identity in $(F, +)$, justify the steps
$$a \cdot 0 + a \cdot b = a \cdot (0 + b)$$
$$= a \cdot b$$
and deduce that $a \cdot 0 = 0$.

5 If $(F, +, \cdot)$ is any field, 0 is the identity in $(F, +)$ and $-b$ is the inverse of b in $(F, +)$, justify the steps
$$a \cdot b + a \cdot - b = a \cdot (b + - b)$$
$$= a \cdot 0$$
$$= 0$$
and deduce that $a \cdot (-b) = -(a \cdot b)$.

6 Prove that $0 \cdot a = 0$ and $(-a) \cdot b = -(a \cdot b)$ with the notation of qn 5.

7 If $a, b \neq 0$ in a field, by considering the multiplicative group, prove that $ab \neq 0$. Deduce that $\mathbb{Z}_4$, constructed by analogy with qns 2 and 3, is *not* a field.

8 Let $(F_4, +, \cdot)$ be a field in which 0 is the identity for $(F, +)$, 1 is the identity for $(F - \{0\}, \cdot)$, and $F_4 = \{0, 1, a, b\}$.
  (i) Make a multiplication table for F_4 using Lagrange's theorem and deduce that $a^2 = b$ and $b^2 = a$.
  (ii) Deduce from qns 5 and 6.7 that $(-1)^2 = 1$ and hence $-1 \neq a, b$.
  (ii) Since $-1 = 1$, show that $a + a = 0$ and $b + b = 0$ and make an addition table for F_4.

9 An injection of the direct product $\mathbb{Q} \times \mathbb{Q}$ into the real numbers is given by
$$(a, b) \mapsto a + b\sqrt{2}.$$

The set of images under this mapping is denoted by $\mathbb{Q}[\sqrt{2}]$.
Is this mapping an isomorphism of the direct sum $\mathbb{Q} \oplus \mathbb{Q}$
(see qn 10.15) onto $(\mathbb{Q}[\sqrt{2}], +)$? Prove that $\mathbb{Q}[\sqrt{2}]$ is closed under
multiplication and that $a + b\sqrt{2}$ has a multiplicative inverse in
this set provided a and b are not both 0. Is $(\mathbb{Q}[\sqrt{2}], +, \cdot)$ a field?

10 If $(F, +, \cdot)$ is a field and addition on the set $F \times F$ is defined by
$(x_1, y_1) + (x_2, y_2) = (x_1 + x_2, y_1 + y_2)$ and multiplication defined
by $(x_1, y_1)(x_2, y_2) = (x_1 x_2, y_1 y_2)$, show that $F \times F$ with these
operations cannot be a field by choosing nonzero elements whose
product is $(0, 0)$ thus contradicting qn 7.

Vector spaces

11 If F is a field, then the operation of *vector addition* on
$F^n = F \times F \times F \times \ldots \times F$ is

$(a_1, a_2, a_3, \ldots, a_n) + (b_1, b_2, b_3, \ldots, b_n)$

$\quad = (a_1 + b_1, a_2 + b_2, a_3 + b_3, \ldots, a_n + b_n)$.

Show that F^n is a group under vector addition.
The identity in this group is called the *zero vector* and written **0**.

12 If F is a field, which of the following subsets of F^2 are subgroups under
vector addition?
  (i) $\{(x, 0)|x \in F\}$,
  (ii) $\{(x, 1)|x \in F\}$,
  (iii) $\{(x, 2x)|x \in F\}$,
  (iv) $\{(x, y)|x + y = 0\}$,
  (v) $\{(x, y)|2x + 3y = 0\}$,
  (vi) $\{(x, y)|x + y = 1\}$.

13 Illustrate on a graph the subsets of $\mathbb{R}^2$ such that
  (i) $y = 0$, (ii) $y = 1$, (iii) $2x + 3y = 0$, (iv) $2x + 3y = 1$,
  (v) $(x, y) \in \mathbb{Z}^2$.
Which of these are subgroups of $\mathbb{R}^2$ under vector addition and
which though not subgroups are cosets of subgroups of $\mathbb{R}^2$?

14 Which of the following mappings give isomorphisms of $\mathbb{R}^2$ to $\mathbb{R}^2$ under
vector addition?
  (i) $(x, y) \mapsto (2x, 2y)$,
  (ii) $(x, y) \mapsto (y, x)$,
  (iii) $(x, y) \mapsto (x + 1, y)$
  (iv) $(x, y) \mapsto (x^3, y)$,
  (v) $(x, y) \mapsto (2x, 3y)$,

(vi) $(x, y) \mapsto (2x, 3y) + (y, x)$,

(vii) $(x, y) \mapsto (x, x)$,

(viii) $(x, y) \mapsto (sx, sy)$ (try both $s = 0$ and $s \neq 0$).

15 Let $(F, +, \cdot)$ be a field. The set F^n with the operation of vector addition forms an abelian group. If $\mathbf{v} = (x_1, x_2, \ldots, x_n) \in F^n$ and $a \in F$, we define $a\mathbf{v} = (ax_1, ax_2, \ldots, ax_n) \in F^n$ to be the *scalar multiple* of $\mathbf{v}$ by a. The set F^n with the operations of vector addition and scalar multiplication is said to form the *vector space* $V_n(F)$. Check the following results

  (i) $1\mathbf{v} = \mathbf{v}$ for all $\mathbf{v} \in F^n$,

  (ii) if $\mathbf{0}$ is the identity in the group F^n (the zero vector) then $a\mathbf{0} = \mathbf{0}$ for all $a \in F$,

  (iii) $0\mathbf{v} = \mathbf{0}$ for all $\mathbf{v} \in F^n$,

  (iv) if $a\mathbf{v} = \mathbf{0}$, then either $a = 0$ or $\mathbf{v} = \mathbf{0}$,

  (v) $(ab)\mathbf{v} = a(b\mathbf{v})$,

  (vi) $(a + b)\mathbf{v} = a\mathbf{v} + b\mathbf{v}$,

  (vii) $a(\mathbf{u} + \mathbf{v}) = a\mathbf{u} + a\mathbf{v}$.

16 When a subset of $V_n(F)$ forms a group under vector addition and is closed under scalar multiplication it is called a *subspace* of $V_n(F)$. Which of the following are subspaces of $V_2(\mathbb{R})$

  (i) $\mathbb{Z}^2$,

  (ii) $\{(x, x) | x \in \mathbb{R}\}$,

  (iii) $\{(0, x) | x \in \mathbb{R}\}$,

  (iv) $\{(x, y) | ax + by = 1\}$?

Why is it sufficient to check for closure under vector addition and scalar multiplication?

17 Which of the following are subspaces of $V_3(F)$

  (i) $\{(x, y, z) | ax + by + cz = 0\}$,

  (ii) $\{(x, y, z) | ax + by + cz = 1\}$,

  (iii) $\{(x, y, z) | ax = by = cz\}$?

18 If $\mathbf{v}$ is a given vector of $V_n(F)$, prove that the set $\{a\mathbf{v} | a \in F\}$ forms a subspace. This is called the subspace *spanned* by $\mathbf{v}$, $Sp(\mathbf{v})$. A subspace formed by the set of all scalar multiples of a nonzero vector is called a *1-dimensional* subspace.

19 Assuming that a, b and c are nonzero, prove that qn 17(iii) is a 1-dimensional subspace. What about 16(ii) and (iii)?

20 What is the geometrical description of a 1-dimensional subspace of $V_3(\mathbb{R})$?

21 Find all the 1-dimensional subspaces of $V_2(\mathbb{Z}_3)$. Does each nonzero vector lie in a unique 1-dimensional subspace?

22 If **u** and **v** are nonzero vectors and $\mathbf{u} \in Sp(\mathbf{v})$, prove that $\mathbf{v} \in Sp(\mathbf{u})$ and in fact $Sp(\mathbf{u}) = Sp(\mathbf{v})$.

23 If **u** and **v** are vectors in $V_n(F)$, verify that the set $\{a\mathbf{u} + b\mathbf{v} | a, b \in F\}$ forms a subspace of $V_n(F)$. This subspace is called the subspace *spanned* by **u** and **v** and is denoted by $Sp(\mathbf{u}, \mathbf{v})$.

24 If $ad - bc \neq 0$, find scalars l_1 and m_1 such that

$l_1(a, b) + m_1(c, d) = (1, 0)$

and scalars l_2 and m_2 such that

$l_2(a, b) + m_2(c, d) = (0, 1)$.

Deduce that $Sp((a, b), (c, d)) = V_2(F)$.

25 If $ad - bc = 0$, we need to show that $Sp((a, b), (c, d))$ is not the whole of $V_2(F)$.
  (i) Why is this obvious if either (a, b) or $(c, d) = (0, 0)$?
  (ii) If neither (a, b) nor (c, d) is the zero vector, verify that there are nonzero scalars l and m such that $l(a, b) + m(c, d) = (0, 0)$, by considering the possibilities $(l, m) = (d, -b)$ or $(c, -a)$. Thus (c, d) is a scalar multiple of (a, b) and $Sp((a, b), (c, d)) = Sp((a, b))$.

26 If **u**, **v** and **w** are vectors in $V_n(F)$, verify that the set $\{a\mathbf{u} + b\mathbf{v} + c\mathbf{w} | a, b, c \in F\}$ forms a subspace of $V_n(F)$. This subspace is called the subspace *spanned* by **u**, **v** and **w** and is denoted by $Sp(\mathbf{u}, \mathbf{v}, \mathbf{w})$.

27 If $Sp(\mathbf{u}, \mathbf{v}, \mathbf{w}) = V_2(F)$, use qns 24 and 25 to prove that either $Sp(\mathbf{v}, \mathbf{w})$ or $Sp(\mathbf{u}, \mathbf{w})$ or $Sp(\mathbf{u}, \mathbf{v}) = V_2(F)$.

If **u** and **v** are nonzero vectors in $V_n(F)$, $Sp(\mathbf{u}, \mathbf{v})$ is called a *2-dimensional* vector space, provided $Sp(\mathbf{u}) \neq Sp(\mathbf{v})$.

28 If $Sp(\mathbf{u}, \mathbf{v})$ is a 2-dimensional subspace of $V_n(F)$, prove that the direct product of the groups $Sp(\mathbf{u})$ and $Sp(\mathbf{v})$ is isomorphic to $Sp(\mathbf{u}, \mathbf{v})$ as a group of vectors under addition as in qn 10.11.

29 Is qn 17(i) a 2-dimensional subspace whatever the values of a, b and c?

Summary

Definition $(F, +, \cdot)$ is said to be a *field* when
   qn 1   (i) $(F, +)$ is an abelian group with identity 0,
           (ii) $(F - \{0\}, \cdot)$ is an abelian group,
           (iii) $a \cdot (b + c) = a \cdot b + a \cdot c$ and $(a + b) \cdot c = a \cdot c + b \cdot c$ for all $a, b, c \in F$.

Definition   The cartesian product $F^n = F \times F \times \ldots \times F$ is said
   qn 15   to form the vector space $V_n(F)$, and its elements are referred to as vectors when it is furnished with two operations as follows:
           (i) the operation of *vector addition* defined so that
$$(x_1, x_2, \ldots, x_n) + (y_1, y_2, \ldots, y_n)$$
$$= (x_1 + y_1, x_2 + y_2, \ldots, x_n + y_n),$$
           (ii) the operation of *scalar muliplication* is the product of a field element (a scalar) and a vector defined by
$$a(x_1, x_2, \ldots, x_n) = (ax_1, ax_2, \ldots, ax_n).$$

Definition   If a subset of $V_n(F)$ forms a group under vector addi-
   qn 16   tion and is closed under scalar multiplication, then it is said to form a *subspace*.

Definition   The subset of $V_n(F)$ which consists of all scalar multi-
   qn 18   ples of one nonzero vector **v** forms a subspace $Sp(\mathbf{v})$ and such a subspace is said to be *1-dimensional*.

Definition   When **u** and **v** are vectors in $V_n(F)$, the subspace
   qn 23 $\{a\mathbf{u} + b\mathbf{v} | a, b \in F\}$ is called the subspace spanned by **u** and **v** and is denoted by $Sp(\mathbf{u}, \mathbf{v})$.

Theorem   The vectors (a, b) and (c, d) span $V_2(F)$ if and only if
qns 24, 25 $ad - bc \neq 0$.

Definition   When **u** and **v** are nonzero vectors, the subspace
   qn 27 $Sp(\mathbf{u}, \mathbf{v})$ is said to be *2-dimensional* provided $Sp(\mathbf{u}) \neq Sp(\mathbf{v})$.

Historical note

The development of the theory of fields took place during the nineteenth century in the context of the attempt to define precisely the solutions of polynomial equations. The fields of rational and real numbers were familiar to Euclid (300 B.C.). The complex number field was developed in the eighteenth century. The prime fields, $\mathbb{Z}_p$, were studied in *Disquisitiones Arithmeticae* by C. F. Gauss (1801). E. Galois constructed algebraic extensions of $\mathbb{Z}_p$ and proved that such fields had cyclic multiplicative groups (1830). Not until 1893 did E. H. Moore establish that every finite field was isomorphic to a Galois extension.

This isomorphism effectively presumes the abstract definition of a field. Stemming from the attempt to prove Fermat's Last Theorem E. E. Kummer (1849) and J. W. R. Dedekind (1870) developed the theory of algebraic extensions of the rational numbers.

Answers to chapter 11

1 $\mathbb{Z} - \{0\}$ is not a group under multiplication.

2 Yes.

3 The two cyclic groups $(\mathbb{Z}_3, +)$ and $(\mathbb{Z}_3 - \{0\}, \cdot)$ are readily identified. $a(b + c) = ab + ac$ is obvious if $a = 0$ or 1. So only $2(b + c) = 2b + 2c$ needs to be checked. $b \mapsto 2b$ matches the two generators of $(\mathbb{Z}_3, +)$.

7 Multiplicative group is closed, so $a, b \neq 0 \Rightarrow ab \neq 0$. In $\mathbb{Z}_4$, $2 \cdot 2 = 0$.

8 (i) From qns 4 and 7, $0 \cdot x = x \cdot 0 = 0$. From Lagrange's theorem, neither a nor b has order 2, so both have order 3.
(ii) $1 + 1 = 0 \Rightarrow a(1 + 1) = 0$.

9 $1/(a + b\sqrt{2}) = (a - b\sqrt{2})/(a^2 - 2b^2)$. Yes, a field.

10 $(x, 0)(0, y) = (0, 0)$.

11 Identity is $(0, 0, \ldots, 0)$. Inverse of $(a_1, a_2, \ldots, a_n)$ is $(-a_1, -a_2, \ldots, -a_n)$.

12 (i), (iii), (iv), (v).

13 Subgroups: (i), (iii), (v). Cosets: (ii), (iv).

14 (i), (ii), (v), (vi), (viii) for $s \neq 0$.

16 (ii) and (iii) only. $0\mathbf{v} = \mathbf{0}. -1\mathbf{v} = -\mathbf{v}$.

17 (i) and (iii).

18 Closure under vector addition follows from qn 15(vi). Closure under scalar multiplication follows from qn 15(v).

19 In 17 (iii), $(x, y, z) = x(1, a/b, a/c)$. In 16 (ii), $(x, x) = x(1, 1)$. In 16 (iii), $(0, x) = x(0, 1)$.

20 A line through the origin.

21 $\{(0, 0), (1, 0), (2, 0)\}, \{(0, 0), (0, 1), (0, 2)\}, \{(0, 0), (1, 1), (2, 2)\}, \{(0, 0), (1, 2), (2, 1)\}$.

22 If $\mathbf{u} = a\mathbf{v}$ then $b\mathbf{u} = (ba)\mathbf{v}$, so $Sp(\mathbf{u}) \subseteq Sp(\mathbf{v})$. Since $a \neq 0$, $\mathbf{v} = a^{-1}\mathbf{u}$ and $b\mathbf{v} = (ba^{-1})\mathbf{u}$, so $Sp(\mathbf{v}) \subseteq Sp(\mathbf{u})$.

23 Closed under vector addition by qn 15 (vi). Closed under scalar multiplication by qns 15(vii) and (v).

24 $l_1 = d/(ad - bc), m_1 = -b/(ad - bc), l_2 = -c/(ad - bc), m_2 = a/(ad - bc)$. $(x, y) = x(1, 0) + y(0, 1)$.

26 Argue as in qn 23.

27 Let $\mathbf{u} = (u_1, u_2)$, $\mathbf{v} = (v_1, v_2)$, $\mathbf{w} = (w_1, w_2)$. then $u_1v_2 - u_2v_1$, $u_1w_2 - u_2w_1$ and $v_1w_2 - v_2w_1$ cannot all be 0 or else $Sp(\mathbf{u}, \mathbf{v}, \mathbf{w})$ would be at most 1-dimensional. If the first is nonzero then $Sp(\mathbf{u}, \mathbf{v}) = V_2(F) = Sp(\mathbf{u}, \mathbf{v}, \mathbf{w})$.

28 To prove qn 10.11 (i): $a_1\mathbf{u} + b_1\mathbf{v} = a_2\mathbf{u} + b_2\mathbf{v} \Rightarrow (a_1 - a_2)\mathbf{u} = (b_2 - b_1)\mathbf{v}$, but $Sp(\mathbf{u}) \neq Sp(\mathbf{v})$ so $Sp(\mathbf{u}) \cap Sp(\mathbf{v}) = \{\mathbf{0}\}$ from qn 22 and $a_1 = a_2$, $b_1 = b_2$. Condition 10.11 (ii) obvious.

29 If $a = b = c = 0$, qn 17(i) $= V_3(F)$. If one of a, b, c is nonzero, say $a \neq 0$, then $x = (-b/a)y + (-c/a)z$ and the subspace is $\{y(-b/a, 1, 0) + z(-c/a, 0, 1)| y, z \in F\}$ which is 2-dimensional.

12

Linear transformations

A linear transformation is a structure-preserving function of one vector space to another vector space over the same field. The structure that must be preserved is that of vector addition and scalar multiplication of which the geometric analogues, when the field is the real numbers, are parallelograms with one vertex at the origin (for vector addition) and lines through the origin (for scalar multiplication). In particular for every linear transformation the image of the zero vector is the zero vector. In this chapter we show that if the linear transformation is from $V_m(F)$ to $V_n(F)$, then it may be represented by multiplication by a unique $m \times n$ matrix. The rows of this matrix are the images of the basis vectors. Conversely, multiplication by any $m \times n$ matrix gives a linear transformation from $V_m(F)$ to $V_n(F)$.

Concurrent reading: Birkhoff and MacLane, chapter 8, section 1.

1 Which of the functions of qn 11.14 preserve scalar multiplication in the sense that if $\mathbf{v} \mapsto \mathbf{v}'$ then $a\mathbf{v} \mapsto a\mathbf{v}'$ for all $a \in \mathbb{R}$?

2 A function $V_m(F) \to V_n(F)$ which preserves the structure of vector addition and the structure of scalar multiplication, in the sense that if

$\mathbf{v} \mapsto \mathbf{v}'$ and $\mathbf{u} \mapsto \mathbf{u}'$ then $\mathbf{v} + \mathbf{u} \mapsto \mathbf{v}' + \mathbf{u}'$ and $a\mathbf{v} \mapsto a\mathbf{v}'$

for all $a \in F$, is called a *linear transformation* of $V_m(F)$. What is the image of the zero vector under any linear transformation?

3 Show that the set of images under a linear transformation form a subspace of the codomain. This subspace is called the *image space* of the transformation.

4 Show that the full subset of the domain with image the zero vector under a linear transformation is a subspace of the domain. This subspace is called the *kernel* of the transformation.

5 Under a linear transformation show that the image of a 1-dimensional subspace of the domain is either a 1-dimensional subspace or the zero vector. How would you describe this result geometrically if the domain and codomain was either $V_2(\mathbb{R})$ or $V_3(\mathbb{R})$?

6 If $\alpha: V_2(F) \to V_2(F)$ is a linear transformation and $\alpha:(1, 0) \mapsto (a, b)$, what is the image of $(x, 0)$ under α? If $\alpha:(0, 1) \mapsto (c, d)$, what is the image of $(0, y)$? Use $(x, 0) + (0, y) = (x, y)$ to deduce the image of (x, y) under α. The conventional expression for this is

$$\alpha:(x, y) \mapsto (x, y)\begin{pmatrix} a & b \\ c & d \end{pmatrix}.$$

Check that every transformation of this form is a linear transformation.

7 If α is a linear transformation of $V_2(\mathbb{R})$ such that

$$\alpha:(x, y) \mapsto (x, y)\begin{pmatrix} 1 & 1 \\ -1 & 2 \end{pmatrix}$$

use some squared paper to illustrate the images of points in $\mathbb{Z}^2$.

8 If $OABC$ is a parallelogram in $\mathbb{R}^2$ what are the possible geometrical configurations of the images of these four points under a linear transformation $V_2(\mathbb{R}) \to V_2(\mathbb{R})$? With the help of qn 5 describe the image of a parallelogram lattice under such a transformation.

9 If $\alpha: V_2(F) \to V_1(F)$ is a linear transformation such that $\alpha:(1, 0) \mapsto a$ and $\alpha:(0, 1) \mapsto b$, prove that

$\alpha:(x, y) \mapsto xa + yb$.

$xa + yb$ is also written $(x, y)\begin{pmatrix} a \\ b \end{pmatrix}$.

Check that every transformation of this form is a linear transformation.

10 If $\alpha: V_3(F) \to V_2(F)$ is a linear transformation such that

$\alpha:(1, 0, 0) \mapsto (a, b)$,

$\alpha:(0, 1, 0) \mapsto (c, d)$, and

$\alpha:(0, 0, 1) \mapsto (p, q)$,

find the image of (x, y, z) under α. We conventionally write

$$\alpha:(x, y, z) \mapsto (x, y, z)\begin{pmatrix} a & b \\ c & d \\ p & q \end{pmatrix}.$$

11 If $\alpha: V_2(F) \to V_3(F)$ is a linear transformation such that $\alpha:(1, 0) \mapsto (a, b, c)$, and

$\alpha:(0, 1) \mapsto (p, q, r)$,

find the image of (x, y) under α. We conventionally write this

$$\alpha:(x, y) \mapsto (x, y) \begin{pmatrix} a & b & c \\ p & q & r \end{pmatrix}.$$

12 If $\alpha: V_m(F) \to V_n(F)$ is a linear transformation such that under α

$(1, 0, 0, \ldots, 0) \mapsto (a_{11}, a_{12}, a_{13}, \ldots, a_{1n})$,

$(0, 1, 0, \ldots, 0) \mapsto (a_{21}, a_{22}, a_{23}, \ldots, a_{2n})$,

$\qquad \vdots \qquad\qquad\qquad \vdots$

$(0, 0, 0, \ldots, 1) \mapsto (a_{m1}, a_{m2}, a_{m3}, \ldots, a_{mn})$,

then we write

$$\alpha : (x_1, x_2, \ldots, x_m) \mapsto (x_1, x_2, \ldots, x_m) \begin{pmatrix} a_{11} & a_{12} & a_{13} & \cdots & a_{1n} \\ a_{21} & a_{22} & a_{23} & \cdots & a_{2n} \\ & & \vdots & & \\ a_{m1} & a_{m2} & a_{m3} & \cdots & a_{mn} \end{pmatrix}.$$

The rectangular array of a_{ij}s is called a *matrix*, strictly an $m \times n$ matrix, and the definition above defines a product of a row vector (an m-tuple) and an $m \times n$ matrix. How could you describe the result of the product of an m-tuple and an $m \times n$ matrix?

13 Must every transformation of the kind exhibited in qn 12 be a linear transformation?

Summary

Definition   A function $\alpha: V_m(F) \to V_n(F)$ is said to be a linear trans-
qn 2   formation when $(\mathbf{u} + \mathbf{v})\alpha = \mathbf{u}\alpha + \mathbf{v}\alpha$ and $\lambda(\mathbf{u}\alpha) = (\lambda\mathbf{u})\alpha$, for all vectors $\mathbf{u}, \mathbf{v}$ of the domain, and all scalars λ in F.

Theorem   The *image space* of a linear transformation is a sub-
qn 3   space of the codomain.

Theorem   The *kernel* of a linear transformation (the subset map-
qn 4   ped to the zero vector) is a subspace of the domain.

Theorem   Every linear transformation $V_m(F) \to V_n(F)$ is uniquely
qn 12   defined by the images of the basis vectors $(1, 0, 0, \ldots, 0), (0, 1, 0, \ldots, 0)$, etc. of the domain, and can be expressed as $\mathbf{v} \mapsto \mathbf{v}A$, where the images of the basis vectors are the rows of the matrix A.

Historical note

Linear transformations were first studied as 'substitutions' and applied to quadratic forms in number theory during the eighteenth century. The substitution of $(ax + by, cx + dy)$ for (x, y) transforms $x^2 + y^2$ to $(ax + by)^2 + (cx + dy)^2$ and interest focussed on those substitutions for which the two forms took on precisely the same set of integer values. The matrix notation was developed by A. Cayley (1858) as he applied substitutions to homogeneous polynomial expressions in n variables (which he called quantics) of which the quadratic forms of Lagrange and Gauss were a special case.

Answers to chapter 12

1 (i), (ii), (v), (vi), (vii), (viii).

2 If $O \mapsto O'$, then $v = v + O \mapsto v' + O' = v'$, so O' is the zero vector.

3 If $v \mapsto v'$ and $u \mapsto u'$, then $v + u \mapsto v' + u'$ so the set of images is closed under vector addition. Also $av \mapsto av'$, so the set of images is closed under scalar multiplication and the images form a subspace.

4 $v \mapsto O$ and $u \mapsto O \Rightarrow v + u \mapsto O + O = O$ and $av \mapsto aO = O$.

5 If $u \mapsto u'$ then $Sp(u) \to Sp(u')$. If $u' = O$ then $Sp(u') = \{O\}$. A line through the origin is either mapped to a line through the origin or to the origin itself.

6 $\alpha{:}(x, 0) \mapsto (xa, xb)$. $\alpha{:}(0, y) \mapsto (yc, yd)$, so $\alpha{:}(x, y) \mapsto (xa + yc, xb + yd)$.

8 Either a parallelogram or four collinear points.
Either another parallelogram lattice or a set of collinear points.

12 An n-tuple.

13

The general linear group GL(2, F)

This chapter is about those linear transformations which are permutations of a 2-dimensional vector space, sometimes $V_2(F)$ and sometimes $V_2(\mathbb{R})$. The work with $V_2(F)$ develops the general theory. When $V_2(\mathbb{R})$ is under discussion, diagrams will represent our ideas faithfully and the full range of geometrical language is at our disposal.

Concurrent reading: Birkhoff and MacLane, chapter 8, sections 3, 4 and 6; Burn (1977).

Singular and nonsingular transformations

The first five questions of this chapter determine when the linear transformation

$$(x, y) \mapsto (x, y)\begin{pmatrix} a & b \\ c & d \end{pmatrix}$$

belongs to the group $S_{V_2(F)}$. When this happens, such transformations are said to be nonsingular. The rest of the chapter is only concerned with the set of nonsingular linear transformations which forms the general linear group $GL(2, F)$.

1   If $(0, 0)$, (a, b) and (c, d) are noncollinear points of $\mathbb{R}^2$, find the area of the parallelogram with vertices at these three points and $(a + c, b + d)$.

2   If $(x, y) \mapsto (x, y)\begin{pmatrix} a & b \\ c & d \end{pmatrix}$ is a linear transformation of $V_2(\mathbb{R})$, what is the area of the image of the unit square $(0, 0)$, $(1, 0)$, $(0, 1)$ and $(1, 1)$ under this transformation?

3   If $\alpha : (x, y) \mapsto (x, y)\begin{pmatrix} a & b \\ c & d \end{pmatrix}$ is a linear transformation of $V_2(F)$,

show that the image space of α is $Sp((a, b), (c, d))$. Deduce from qn 11.25 that if $ad - bc = 0$ then α is not a surjection onto $V_2(F)$. By considering the images of the vectors $(0, 0)$, $(d, -b)$ and $(-c, a)$ show that if $ad - bc = 0$, then α is not one–one.

The number $ad - bc$ is called the *determinant* of the matrix. When the determinant of a matrix is 0, the matrix and its related linear transformation are said to be *singular*.

4 If $ad - bc \neq 0$, so that the linear transformation α of qn 3 is *non-singular*, use qn 11.24 to show that α is a surjection $V_2(F) \rightarrow V_2(F)$. If it was the case that $(x_1, y_1)\alpha = (x_2, y_2)\alpha$, show that $(x_1 - x_2)(a, b) = (y_2 - y_1)(c, d)$, and deduce that α is an injection.

5 Use qns 3 and 4 to formulate a theorem about surjections and injections which are linear transformations $V_2(F) \rightarrow V_2(F)$.

The group of nonsingular transformations

In qns 6–14, matrix multiplication is defined so as to be compatible with the composition of linear transformations and the group of non-singular linear transformations is identified. The isomorphism between matrices and linear transformations is so well used in ordinary mathematical language that the two are often, and quite helpfully, confused.

6 If we denote the matrix $\begin{pmatrix} a & b \\ c & d \end{pmatrix}$ by A and the matrix $\begin{pmatrix} p & q \\ r & s \end{pmatrix}$ by P and we let α denote the linear transformation $(x, y) \mapsto (x, y)A$ and let β denote the linear transformation $(x, y) \mapsto (x, y)P$, then the composite linear transformation $\alpha\beta$ is given by $(x, y) \mapsto [(x, y)A]P$. Find the matrix corresponding to the linear transformation $\alpha\beta$. This matrix is defined to be the *matrix product AP*.

7 If I is the matrix $\begin{pmatrix} 1 & 0 \\ 0 & 1 \end{pmatrix}$, how could you describe the linear transformation $(x, y) \mapsto (x, y)I$?

8 If $A = \begin{pmatrix} a & b \\ c & d \end{pmatrix}$ is a nonsingular matrix with determinant Δ, find the product of A with $\begin{pmatrix} d/\Delta & -b/\Delta \\ -c/\Delta & a/\Delta \end{pmatrix}$. Deduce that the linear transformation $(x, y) \mapsto (x, y)A$ has an inverse which is a linear transformation.

9 If A and B are 2×2 matrices over the same field prove that the determinant of $AB = $ (det of A) $\cdot$ (det of B).

10 If A is a 2×2 matrix, and if A has an inverse with respect to the identity matrix $\begin{pmatrix} 1 & 0 \\ 0 & 1 \end{pmatrix}$, prove that the determinant of $A \neq 0$.

11 Use the definition of matrix multiplication and the associativity of composible functions (qn 1.32) to prove that products of 2×2 matrices are associative.

12 Prove that the set of 2×2 nonsingular matrices over a given field forms a group under matrix multiplication.

13 If V is a vector space, and $\alpha: V \to V$ is a linear transformation which is also a bijection, then α is called a nonsingular linear transformation of V. Must the nonsingular linear transformations of V form a subgroup of S_V?

14 Prove that the group of linear bijections $V_2(F) \to V_2(F)$ is isomorphic to the group of 2×2 nonsingular matrices over F. Either one of these groups is denoted by $GL(2, F)$ and called the *general linear group* of $V_2(F)$.

The centre of the general linear group

In questions 15–19, the elements of $GL(2, F)$ which commute with all the rest of the group are identified and are shown to form a subgroup.

15 A matrix of the form $\begin{pmatrix} a & 0 \\ 0 & a \end{pmatrix}$ in $GL(2, F)$ is sometimes denoted by aI and called a *scalar* matrix.
Under the isomorphism of qn 14 what linear transformations of $V_2(\mathbb{R})$ correspond to scalar matrices?

16 Let $A = \begin{pmatrix} a & b \\ c & d \end{pmatrix}$, $J = \begin{pmatrix} 1 & 1 \\ 0 & 1 \end{pmatrix}$ and $K = \begin{pmatrix} 1 & 0 \\ 1 & 1 \end{pmatrix}$. If $AJ = JA$, prove that $a = d$ and $c = 0$. If $AK = KA$, prove that $a = d$ and $b = 0$.
Deduce that if A is in $GL(2, F)$ and A commutes with every matrix in $GL(2, F)$ then A is a scalar matrix. Check that scalar matrices always commute with all other matrices of $GL(2, F)$.

17 Does the subset of scalar matrices form a subgroup of $GL(2, F)$?

18 In any group G, prove that the subset $C = \{c \mid cg = gc \text{ for all } g \in G\}$ forms a subgroup of G.
This subgroup is called the *centre* of G.

19 Identify the centres of D_2, D_3 and D_4.

Shears

In questions 20–24, some shears of $V_2(\mathbb{R})$ are identified, and a general definition of a shear in $V_n(F)$ is given. One example is given of the isomorphism of the additive group of the field with the group of shears with a given axis.

20 Under the isomorphism of qn 14, illustrate the linear transformation of $V_2(\mathbb{R})$ which corresponds to the matrix $\begin{pmatrix} 1 & 0 \\ 1 & 1 \end{pmatrix}$. This transformation is called a *shear* on the *x*-axis. A shear of $V_2(\mathbb{R})$ has a line of fixed points and every point off this line is moved parallel to this line.

21 Find the line of fixed points of the shear

$$(x, y) \mapsto (x, y)\begin{pmatrix} 1 & a \\ 0 & 1 \end{pmatrix}.$$

22 Find the line of fixed points of each of the shears

   (i) $(x, y) \mapsto (x, y) + y(a, 0)$,   where $a \neq 0$,

   (ii) $(x, y) \mapsto (x, y) + x(0, b)$,   where $b \neq 0$,

   (iii) $(x, y) \mapsto (x, y) + (cx + dy)(a, b)$,

   where (a, b) is not the zero vector and $ca + db = 0$.

23 If $\mathbf{v} \mapsto \mathbf{v} + f(\mathbf{v})\mathbf{a}$, where $f(\mathbf{v})$ is a scalar and $\mathbf{a}$ a constant vector, is a linear transformation of $V_n(F)$, prove that the function f is a linear transformation $V_n(F) \to F$. When $f(\mathbf{a}) = 0$, the form given here is the general definition of a *shear* (or *transvection*) of a general vector space. Show that each of the shears in qn 22 may be expressed in the form given here by finding the function f and the constant vector $\mathbf{a}$ in each case.

24 Show that the matrices of the type $\begin{pmatrix} 1 & 0 \\ a & 1 \end{pmatrix}$ form a subgroup of $GL(2, F)$ isomorphic to $(F, +)$.

More subgroups of *GL*(2, *F*)

25 Show that the nonsingular matrices of the type $\begin{pmatrix} a & 0 \\ 0 & 1 \end{pmatrix}$ form a subgroup of $GL(2, F)$ isomorphic to the multiplicative group $F\text{-}\{0\}$. What are the fixed points and fixed lines of the corresponding group of linear transformations of $V_2(\mathbb{R})$?

26 How would your description of the fixed points and fixed lines in qn 25 be affected if the real numbers were replaced by an arbitrary field and you were confined to the language of vector spaces and groups (e.g. if lines through the origin always had to be referred to as 1-dimensional subspaces)?

27 Show that the nonsingular matrices of the type $\begin{pmatrix} a & 0 \\ 0 & b \end{pmatrix}$ form a group. What are the fixed points and fixed lines of the corresponding group of linear transformations of $V_2(\mathbb{R})$?
Matrices of this type are called *diagonal* matrices.

28 If $G = (\mathbb{R}\text{-}\{0\}, \cdot)$ show that the group of diagonal matrices in $GL(2, \mathbb{R})$ is isomorphic to the direct product $G \times G$.

29 Modify your description of the fixed points and fixed lines in qn 27 so that it would be valid over any field.

30 Show that the matrices of the type $\begin{pmatrix} a & 0 \\ c & d \end{pmatrix}$, where $a, d \neq 0$, form a group under matrix multiplication. What are the fixed lines and fixed families of lines of the corresponding group of linear transformations of $V_2(\mathbb{R})$? Matrices of this type are called (lower) *triangular* matrices.

31 How would your description of the fixed lines in qn 30 be affected if the real numbers were replaced by an arbitrary field?

32 In $GL(2, \mathbb{Z}_3)$ find the orders of the group of scalar matrices, the group of shears on the x-axis, the group of diagonal matrices and the group of triangular matrices. Determine the order of the full group by counting the possible pairs of images of $(1, 0)$ and $(0, 1)$. There are nine vectors in $V_2(\mathbb{Z}_3)$. How many of these may be images for $(1, 0)$? After determining the image for $(1, 0)$, how many may be images for $(0, 1)$?

The orthogonal group

In chapter 21 we will link the orthogonal group with isometries in $GL(2, \mathbb{R})$ and $GL(3, \mathbb{R})$, but here we simply offer a method of defining a subgroup of $GL(2, F)$ which is quite different from any of those we have used so far.

33 If $A = \begin{pmatrix} a & b \\ c & d \end{pmatrix}$, we define the *transpose* of A, $A^\mathrm{T} = \begin{pmatrix} a & c \\ b & d \end{pmatrix}$. Prove that $\det A = \det A^\mathrm{T}$ and for any two 2×2 matrices A and B that $(AB)^\mathrm{T} = B^\mathrm{T} A^\mathrm{T}$.

34 Prove that the 2×2 matrices A such that $A \cdot A^\mathrm{T} = I$ form a subgroup of $GL(2, F)$. This subgroup is known as the *orthogonal* group. Prove that if $A = \begin{pmatrix} a & b \\ c & d \end{pmatrix}$ then $a^2 + b^2 = c^2 + d^2 = 1$, $ac + bd = ab + cd = 0$ and $ad - bc = \pm 1$.

35 If $\begin{pmatrix} a & b \\ c & d \end{pmatrix}$ is in the orthogonal subgroup of $GL(2, \mathbb{R})$ and if $a = \cos \theta$, $d = \cos \phi$, prove that $\theta \pm \phi = 0$ or π, and exhibit the possible forms of the matrices in this orthogonal group.

Summary

Definition The *determinant* of the 2×2 matrix $\begin{pmatrix} a & b \\ c & d \end{pmatrix}$ is $ad - bc$.
    qn 3

Definition The linear transformation $\mathbf{v} \mapsto \mathbf{v}A$ of $V_2(F)$ and the
qn 3 2×2 matrix A are said to be *singular* when $\det A = 0$.

Theorem   A singular linear transformation of $V_2(F)$ is neither
qn 3   one–one nor onto.

Definition The linear transformation $\mathbf{v} \mapsto \mathbf{v}A$ of $V_2(F)$ and the
qn 4 2×2 matrix A are said to be *nonsingular* when
         $\det A \neq 0$.

Theorem   A nonsingular linear transformation is a bijection.
qn 4

Definition If A and B are 2×2 matrices, the *matrix product* AB is
qn 6   the matrix of the composite transformation
         $[\mathbf{v} \mapsto \mathbf{v}A$ followed by $\mathbf{v} \mapsto \mathbf{v}B]$.

Theorem   If A and B are 2×2 matrices, then $\det AB =$
qn 9 $\det A \cdot \det B$.

Definition The nonsingular linear transformations of $V_2(F)$ form
qns 13, 14 the *general linear group* $GL(2, F)$.

Theorem   The group $GL(2, F)$ is isomorphic to the group of
qn 14 2×2 nonsingular matrices over the field F.

Definition The set of elements in a group which commute with
qn 18   every element in that group is called the *centre* of the
         group.

Theorem   The centre of a group forms a subgroup.
qn 18

Definition A square matrix is said to be a *scalar* matrix if all the
qn 15   entries on the main diagonal are the same and the
         remaining entries are zero.

Theorem   The subset of scalar matrices is the centre of the general
qn 16   linear group.

Definition When the only nonzero entries in a square matrix lie
qn 27   down the main diagonal, the matrix is said to be
         diagonal.

Definition The *transpose* A^T of a matrix A is found by interchang-
qn 33   ing the rows and columns of A.

Theorem   The subset of matrices A such that $A \cdot A^T = I$ is a sub-
qn 34   group of the general linear group called the *orthogonal*
         group.

Historical note

As we said at the beginning of chapter 12, linear transformations
were first studied, without the use of matrices, as substitutions applied
to quadratic forms. In Gauss' *Disquisitiones Arithmeticae* (1801) the
determinant $ad - bc$ of the substitution of $(ax + by, cx + dy)$ for
(x, y) appeared since the discriminant $B^2 - AC$ of the quadratic form

$Ax^2 + 2Bxy + Cy^2$ is multiplied by the square of the determinant by the substitution.

The use of a rectangular array for a matrix is due to Cayley, who, in a paper read to the Royal Society in 1858, gave many of the rules for the algebra of 2×2 and 3×3 matrices.

The general linear group in two dimensions over $\mathbb{Z}_p$ was studied by E. Galois (1830) and the general linear group in n dimensions over $\mathbb{Z}_p$ was discussed fully by C. Jordan in his *Traité des Substitutions* (1870). Neither Galois nor Jordan used matrices in these expositions.

Answers to chapter 13

1 $|ad - bc|$.

2 $|ad - bc|$.

3 $Sp((1, 0), (0, 1)) \to Sp((a, b), (c, d))$.

4 $(x_1 - x_2)(a, b) = (y_2 - y_1)(c, d) \Rightarrow$ either $x_1 = x_2$ and $y_1 = y_2$ or $Sp((a, b), (c, d))$ is 1-dimensional.

5 If $ad - bc \neq 0$ then $\alpha : (x, y) \mapsto (x, y) \begin{pmatrix} a & b \\ c & d \end{pmatrix}$ is a bijection.

If $ad - bc = 0$ then α is neither a surjection nor an injection.

6 $\begin{pmatrix} ap + br & aq + bs \\ cp + dr & cq + ds \end{pmatrix}$.

7 The identity map on $V_2(F)$.

8 Product $= I$.

9 From qn 6, $(ap + br)(cq + ds) - (cp + dr)(aq + bs)$
$$= apds + brcq - cpbs - draq$$
$$= (ad - bc)(ps - rq).$$

10 $A \cdot A^{-1} = I$ implies $\det A \cdot \det A^{-1} = 1$ so $\det A \neq 0$.

12 Closure follows from qn 9. Identity from qn 7. Inverses from qn 8.

13 Composite of two bijections is a bijection. Composite of two linear transformations is linear. Inverse of linear bijection is linear.

14 The function mapping $[(x, y) \mapsto (x, y)A]$ to A is an isomorphism.

15 Enlargements with centre $(0, 0)$.

17 Yes.

19 D_2 is all centre. Centre of D_3 is $\{e\}$. Centre of $D_4 = \{e, a^2\}$ see qn 9.31.

20 Each point of the x-axis is fixed. Each line parallel to the x-axis is mapped onto itself.

21 The y-axis.

22 (i) $y = 0$. (ii) $x = 0$. (iii) $Sp((a, b))$.

23 $\mathbf{u} + f(\mathbf{u})\mathbf{a} + \mathbf{v} + f(\mathbf{v})\mathbf{a} = \mathbf{u} + \mathbf{v} + f(\mathbf{u} + \mathbf{v})\mathbf{a} \Rightarrow f(\mathbf{u} + \mathbf{v}) = f(\mathbf{u}) + f(\mathbf{v})$,
and $k(\mathbf{v} + f(\mathbf{v})\mathbf{a}) = k\mathbf{v} + f(k\mathbf{v})\mathbf{a} \Rightarrow kf(\mathbf{v}) = f(k\mathbf{v})$.
(i) $\mathbf{a} = (a, 0)$. $f(x, y) = y$. (ii) $\mathbf{a} = (0, b)$, $f(x, y) = x$. (iii) $\mathbf{a} = (a, b)$, $f(x, y) = cx + dy$.

24 $\begin{pmatrix} 1 & 0 \\ a & 1 \end{pmatrix} \mapsto a$ gives the isomorphism.

25 The y-axis is fixed pointwise. Lines parallel to the x-axis are fixed but not pointwise.

26 $Sp(0, 1)$ is fixed pointwise. Cosets of $Sp(1, 0)$ are also fixed but not pointwise.

27 Only the x- and y-axis.

28 $\begin{pmatrix} a & 0 \\ 0 & b \end{pmatrix} \mapsto (a, b)$.

29 $Sp(1, 0)$ and $Sp(0, 1)$ are fixed but not pointwise.

30 The x-axis is fixed but not pointwise. Every line parallel to the x-axis is mapped to a parallel line.

31 Refer to $Sp(1, 0)$ and its cosets.

32 |Scalar matrices| = 2, |Shears on x-axis| = 3, |Diagonal matrices| = 4, |Triangular matrices| = 12. Eight possible images for $(1, 0)$, then six possible for $(0, 1)$ so $|GL(2, \mathbb{Z}_3)| = 48$.

34 If $A \cdot A^{\mathsf{T}} = B \cdot B^{\mathsf{T}} = I$ then $AB \cdot (AB)^{\mathsf{T}} = ABB^{\mathsf{T}}A^{\mathsf{T}} = I$, so subset closed. $A^{\mathsf{T}} = A^{-1}$ and $(A^{\mathsf{T}})^{\mathsf{T}} = A$, so inverse is orthogonal. $\det(A \cdot A^{\mathsf{T}}) = \det I$ and $\det A = \det A^{\mathsf{T}} \Rightarrow \det A = \pm 1$. Other equations follow from $A \cdot A^{\mathsf{T}} = I$ or $A^{\mathsf{T}} \cdot A = I$.

35 $b^2 = \sin^2\theta$, $c^2 = \sin^2\phi$, $ac + bd = 0$ gives $\sin(\phi \pm \theta) = 0$. $\theta = \pm\phi \Rightarrow bc = \cos^2\theta - 1 = -\sin^2\theta$. $\theta = \pm\phi + \pi \Rightarrow bc = 1 - \cos^2\theta = \sin^2\theta$.
$$\begin{pmatrix} \cos\theta & \sin\theta \\ -\sin\theta & \cos\theta \end{pmatrix} \text{ and } \begin{pmatrix} \cos\theta & \sin\theta \\ \sin\theta & -\cos\theta \end{pmatrix}.$$

14

The vector space $V_3(F)$

Angles, distances, areas and volumes in $V_3(\mathbb{R})$ can be calculated with the help of two operations on vectors, the scalar and vector products. In combination, these enable us to define the determinant of a 3×3 matrix and to establish that 3×3 determinants have similar properties to 2×2 determinants.

Throughout this chapter, as in preceding chapters, we will think of vectors as points.

Concurrent reading: Birkhoff and MacLane, chapter 7, section 9; Maxwell chapter 13.

Scalar products

1 If $\mathbf{a} = (a_1, a_2, a_3)$ and $\mathbf{b} = (b_1, b_2, b_3)$ are points of $\mathbb{R}^3$ use the cosine formula to prove that
$$\cos \mathbf{a0b} = \frac{a_1 b_1 + a_2 b_2 + a_3 b_3}{\sqrt{(a_1^2 + a_2^2 + a_3^2)(b_1^2 + b_2^2 + b_3^2)}}.$$
Assume that the distance $\mathbf{0a} = \sqrt{(a_1^2 + a_2^2 + a_3^2)}$.

2 With the notation of qn 1, describe the configuration of the points $\mathbf{a}$, $\mathbf{0}$, and $\mathbf{b}$ when $a_1 b_1 + a_2 b_2 + a_3 b_3 = 0$.

3 In $V_3(F)$ we define a *scalar product* (or *inner product*) of two vectors $\mathbf{a} = (a_1, a_2, a_3)$ and $\mathbf{b} = (b_1, b_2, b_3)$ by
$$\mathbf{a} \cdot \mathbf{b} = a_1 b_1 + a_2 b_2 + a_3 b_3.$$
Use the scalar product to give an expression for the length of $\mathbf{0a}$ in case $F = \mathbb{R}$. We will denote this length by $|\mathbf{a}|$. Check that in $V_3(\mathbb{R})$, $\mathbf{a} \cdot \mathbf{b} = |\mathbf{a}| \cdot |\mathbf{b}| \cos \mathbf{a0b}$.

4 For some constant vector $\mathbf{a}$, express the transformation of $V_3(F)$ given by

$$\mathbf{v} \mapsto \mathbf{v} \cdot \mathbf{a}$$

using matrices, and deduce that this is a linear transformation. What is the codomain of this transformation? Use qns 2 and 11.17(i) to describe the kernel of this transformation.

Vector products

5 In $V_3(F)$ we define a *vector product* (or *outer product*) of two vectors $\mathbf{a} = (a_1, a_2, a_3)$ and $\mathbf{b} = (b_1, b_2, b_3)$ by

$$\mathbf{a} \times \mathbf{b} = \left(\det \begin{pmatrix} a_2 & a_3 \\ b_2 & b_3 \end{pmatrix}, \ \det \begin{pmatrix} a_3 & a_1 \\ b_3 & b_1 \end{pmatrix}, \ \det \begin{pmatrix} a_1 & a_2 \\ b_1 & b_2 \end{pmatrix} \right).$$

Check that $\mathbf{b} \times \mathbf{a} = -\mathbf{a} \times \mathbf{b}$. In $V_3(\mathbb{R})$ check that $|\mathbf{a} \times \mathbf{b}| = |\mathbf{a}| \cdot |\mathbf{b}| \sin \mathbf{a0b}$, and deduce that $|\mathbf{a} \times \mathbf{b}|$ is the area of the parallelogram with vertices $\mathbf{0}, \mathbf{a}, \mathbf{b}, \mathbf{a} + \mathbf{b}$.

6 For some constant vector $\mathbf{a}$, express the transformation of $V_3(F)$ given by

$$\alpha : \mathbf{v} \mapsto \mathbf{v} \times \mathbf{a}$$

using matrices, and deduce that this is a linear transformation. Find $\mathbf{a} \times \mathbf{a}$ and deduce that $Sp(\mathbf{a})$ lies in the kernel of α. Find the kernel of α when $\mathbf{a} = (a_1, 0, 0)$ with $a_1 \neq 0$. Find the kernel of α when $\mathbf{a} = (a_1, a_2, 0)$ with $a_1, a_2 \neq 0$. Find the kernel of α when $\mathbf{a} = (a_1, a_2, a_3)$ with $a_1, a_2, a_3 \neq 0$. Deduce that when $\mathbf{a} \neq \mathbf{0}$, the kernel of α is $Sp(\mathbf{a})$.

7 In $V_3(F)$ find $\mathbf{a} \cdot (\mathbf{a} \times \mathbf{b})$ and $\mathbf{b} \cdot (\mathbf{a} \times \mathbf{b})$ and prove that $Sp(\mathbf{a}, \mathbf{b})$ lies in the kernel of the linear transformation

$$\mathbf{v} \mapsto \mathbf{v} \cdot (\mathbf{a} \times \mathbf{b}).$$

What can you say about the directions of the lines joining the origin to $\mathbf{a}, \mathbf{b}$ and $\mathbf{a} \times \mathbf{b}$? What can you say about the direction of the line joining the origin to $\mathbf{a} \times \mathbf{b}$ and the plane $Sp(\mathbf{a}, \mathbf{b})$? Explain why $(l\mathbf{a} + m\mathbf{b} + n\mathbf{c}) \cdot (\mathbf{b} \times \mathbf{c}) = l\mathbf{a} \cdot (\mathbf{b} \times \mathbf{c})$.

8 Express the transformation of $V_3(F)$ given by

$$\mathbf{v} \mapsto \mathbf{a} \cdot (\mathbf{v} \times \mathbf{c})$$

as a composite of two linear transformations, and deduce that this too is a linear transformation. What is its codomain? Does $\mathbf{a}$ lie in the kernel? Does $\mathbf{c}$ lie in the kernel? Does the whole of $Sp(\mathbf{a}, \mathbf{c})$ lie in the kernel? Explain why $\mathbf{a} \cdot [(l\mathbf{a} + m\mathbf{b} + n\mathbf{c}) \times \mathbf{c}] = m\mathbf{a} \cdot (\mathbf{b} \times \mathbf{c})$.

9 Express the transformation of $V_3(F)$ given by

$$\mathbf{v} \mapsto \mathbf{a} \cdot (\mathbf{b} \times \mathbf{v})$$

as a composite of two linear transformations and deduce that this

too is a linear transformation. What is its codomain? Does $Sp(\mathbf{a}, \mathbf{b})$
lie in its kernel? Explain why $\mathbf{a} \cdot [\mathbf{b} \times (l\mathbf{a} + m\mathbf{b} + n\mathbf{c})] = n\mathbf{a} \cdot (\mathbf{b} \times \mathbf{c})$.

Determinants

10 In the rest of this chapter A will denote the matrix

$$\begin{pmatrix} a_1 & a_2 & a_3 \\ b_1 & b_2 & b_3 \\ c_1 & c_2 & c_3 \end{pmatrix}$$

with rows $\mathbf{a}$, $\mathbf{b}$ and $\mathbf{c}$. Sometimes A will be written in the form

$$\begin{pmatrix} \mathbf{a} \\ \mathbf{b} \\ \mathbf{c} \end{pmatrix}.$$

In $V_3(\mathbb{R})$, calculate the volume of the parallelepiped with vertices $\mathbf{0}$, $\mathbf{a}, \mathbf{b}, \mathbf{c}, \mathbf{b} + \mathbf{c}, \mathbf{c} + \mathbf{a}, \mathbf{a} + \mathbf{b}, \mathbf{a} + \mathbf{b} + \mathbf{c}$, by first using qn 5 to show that the area of the parallelogram with vertices $\mathbf{0}, \mathbf{b}, \mathbf{c}, \mathbf{b} + \mathbf{c}$ is the length of $\mathbf{0}(\mathbf{b} \times \mathbf{c})$ and that this line is perpendicular to the plane of the parallelogram. Then find the volume by multiplying this area by the distance from $\mathbf{a}$ to the plane of the parallelogram. The volume is in fact $\mathbf{a} \cdot (\mathbf{b} \times \mathbf{c})$. Interpret the kernels of the transformations in qns 7, 8 and 9 in the light of this result. Over any field, $\mathbf{a} \cdot (\mathbf{b} \times \mathbf{c})$ is called the *determinant* of A.

11 Use 7, 8 and 9 to find the values of the determinants of the following matrices.

$$\begin{pmatrix} 1 & 2 & 3 \\ 1 & 2 & 3 \\ 4 & 5 & 6 \end{pmatrix}, \quad \begin{pmatrix} 1 & 2 & 3 \\ 4 & 5 & 6 \\ 1 & 2 & 3 \end{pmatrix}, \quad \begin{pmatrix} 1 & 2 & 3 \\ 2 & 4 & 6 \\ 7 & 8 & 9 \end{pmatrix}, \quad \begin{pmatrix} 1 & 2 & 3 \\ 4 & 5 & 6 \\ 5 & 7 & 9 \end{pmatrix}.$$

12 The terms in the expansion of $\mathbf{a} \cdot \mathbf{b} \times \mathbf{c}$ each have the form $\pm a_i b_j c_k$ where $\{i, j, k\} = \{1, 2, 3\}$. Determine exactly which triples (i, j, k) have a positive sign attached and which a negative sign. If the triple (i, j, k) is associated with the permutation $\begin{pmatrix} 1 & 2 & 3 \\ i & j & k \end{pmatrix}$, name the set of permutations which gives a positive coefficient in the expression for the determinant and the set which gives a negative coefficient.

13 What is the relationship between

$\mathbf{a} \cdot \mathbf{b} \times \mathbf{c}, \mathbf{a} \cdot \mathbf{c} \times \mathbf{b}, \mathbf{b} \cdot \mathbf{c} \times \mathbf{a}, \mathbf{b} \cdot \mathbf{a} \times \mathbf{c}, \mathbf{c} \cdot \mathbf{a} \times \mathbf{b}$ and $\mathbf{c} \cdot \mathbf{b} \times \mathbf{a}$?

14 If B is the matrix with columns $\mathbf{b} \times \mathbf{c}, \mathbf{c} \times \mathbf{a}$ and $\mathbf{a} \times \mathbf{b}$, use the principles of qn 13.6 to evaluate the matrix product AB.

15 Use the working of qn 14 to find a 3 × 3 matrix inverse to

(i) $\begin{pmatrix} 1 & 0 & 1 \\ 1 & 1 & 1 \\ 1 & 1 & 0 \end{pmatrix}$,   (ii) $\begin{pmatrix} 1 & 2 & 3 \\ 2 & 3 & 0 \\ 0 & 1 & 2 \end{pmatrix}$,   (iii) $\begin{pmatrix} 2 & -1 & 0 \\ 1 & 2 & 1 \\ -1 & 0 & 3 \end{pmatrix}$.

Check your result in each case.

16 Deduce from qn 14 that if det $A \neq 0$, then the matrix A and the linear transformation $\mathbf{v} \mapsto \mathbf{v}A$ are both invertible.

17 Find $(a_1, b_1, c_1) \cdot (a_2, b_2, c_2) \times (a_3, b_3, c_3)$ and deduce that det $A = $ det A^T. Use $(AB)^T = B^T A^T$ and qn 14 to show that if det $A \neq 0$, then A has both a left and a right inverse.

18 If U is the matrix

$\begin{pmatrix} u_1 & u_2 & u_3 \\ v_1 & v_2 & v_3 \\ w_1 & w_2 & w_3 \end{pmatrix}$

with the vectors $\mathbf{u}$, $\mathbf{v}$ and $\mathbf{w}$ as its rows, verify that the rows of the matrix product UA are

$u_1\mathbf{a} + u_2\mathbf{b} + u_3\mathbf{c}, v_1\mathbf{a} + v_2\mathbf{b} + v_3\mathbf{c}$ and $w_1\mathbf{a} + w_2\mathbf{b} + w_3\mathbf{c}$.

19 Use qns 18 and 7, 8 and 9 to prove that the determinant of

$$UA = u_1\mathbf{a} \cdot [v_2\mathbf{b} \times w_3\mathbf{c} + v_3\mathbf{c} \times w_2\mathbf{b}]$$
$$+ u_2\mathbf{b} \cdot [v_1\mathbf{a} \times w_3\mathbf{c} + v_3\mathbf{c} \times w_1\mathbf{a}]$$
$$+ u_3\mathbf{c} \cdot [v_1\mathbf{a} \times w_2\mathbf{b} + v_2\mathbf{b} \times w_1\mathbf{a}].$$

20 Use qns 19 and 13 to prove that the determinant of $UA = $
$u_1\mathbf{a} \cdot [(v_2w_3 - v_3w_2)\mathbf{b} \times \mathbf{c}] + u_2\mathbf{b} \cdot [(v_3w_1 - v_1w_3)\mathbf{c} \times \mathbf{a}]$
$+ u_3\mathbf{c} \cdot [(v_1w_2 - v_2w_1)\mathbf{a} \times \mathbf{b}].$

21 Use qns 20, 12 and 13 to prove that the determinant of $UA = $
$[\mathbf{u} \cdot \mathbf{v} \times \mathbf{w}][\mathbf{a} \cdot \mathbf{b} \times \mathbf{c}] = $ det $U \cdot$ det A.

22 Deduce from qn 21 that if A is invertible, then det $A \neq 0$.

Singular and nonsingular transformations

23 When det $A \neq 0$, both the transformation $\alpha : \mathbf{v} \mapsto \mathbf{v}A$ and the matrix A are said to be *nonsingular*. Use qns 16 and 17 to show that in this case $\alpha \in S_{V_3(F)}$.

24 Prove that the nonsingular linear transformations of $V_3(F)$ form a group, the nonsingular 3 × 3 matrices form a group under matrix multiplication and that these two groups are isomorphic. Either group is called, abstractly, the *general linear group GL(3, F)*.

25 When det $A = 0$, both the transformation $\alpha : \mathbf{v} \mapsto \mathbf{v}A$ and the matrix A are said to be *singular*. Why is $Sp(\mathbf{a}, \mathbf{b}, \mathbf{c})$ the image space of α? When A is singular, prove that $Sp(\mathbf{a}, \mathbf{b}, \mathbf{c})$ lies in the kernel of the transformation $\mathbf{v} \mapsto \mathbf{v} \cdot \mathbf{b} \times \mathbf{c}$, so that $Sp(\mathbf{a}, \mathbf{b}, \mathbf{c})$ cannot be the whole of $V_3(F)$ and α is not a surjection.

26 When A is singular use qn 14 and $(AB)^T = B^T A^T$ to construct a nonzero vector in the kernel of $\mathbf{v} \mapsto \mathbf{v}A$, so that this transformation is not an injection.

27 Prove that the centre of the group $GL(3, F)$ consists of scalar matrices.

28 Use the definition of a shear in qn 13.23 to show that every shear of $V_3(F)$ has the form

$$(x, y, z) \mapsto (x, y, z) + (px + qy + rz)(a, b, c)$$

and state the further condition which is necessary. Show that in $V_3(F)$ a shear has a 2-dimensional subspace of fixed points.

29 In $GL(3, \mathbb{Z}_2)$ identify the order of the subgroup of scalar matrices, the order of the group of shears fixing a given 2-dimensional subspace, the order of the lower triangular group and the order of the full group.

Summary

Definition   If $\mathbf{a} = (a_1, a_2, a_3)$ and $\mathbf{b} = (b_1, b_2, b_3)$ then the *scalar*
qns 3, 5   *product*

$\mathbf{a} \cdot \mathbf{b} = a_1 b_1 + a_2 b_2 + a_3 b_3 \in F$ and the *vector product*
$\mathbf{a} \times \mathbf{b} = (a_2 b_3 - a_3 b_2, a_3 b_1 - a_1 b_3, a_1 b_2 - a_2 b_1)$.

Definition   If $\mathbf{a}, \mathbf{b}$ and $\mathbf{c}$ are the rows of a square matrix A, then the
qn 10   *determinant* of A is $\mathbf{a} \cdot (\mathbf{b} \times \mathbf{c})$.

Theorem   In $V_3(\mathbb{R})$, the absolute value of the determinant
qn 10 $\mathbf{a} \cdot (\mathbf{b} \times \mathbf{c})$ is the volume of the parallelepiped with three concurrent edges through the origin and the points $\mathbf{a}$, $\mathbf{b}$ and $\mathbf{c}$.

Theorem   If A and B are 3×3 matrices then det $AB =$
qn 21   det $A \cdot$ det B.

Definition   If $\alpha : \mathbf{v} \mapsto \mathbf{v}A$ is a linear transformation of $V_3(F)$, then α
qns 23, 25   and A are said to be nonsingular when det $A \neq 0$ and are said to be singular when det $A = 0$.

Theorem   The nonsingular linear transformations of $V_3(F)$ form a
qn 24   group which is isomorphic to the group of nonsingular 3×3 matrices under matrix multiplication. Either of these two groups is known, abstractly, as the *general linear group* $GL(3, F)$.

Historical note

The notation of scalar product and vector product of two vectors was devised by J. W. Gibbs (1881) in order to simplify the description of 3-dimensional real space which had become available following W. R. Hamilton's discovery of the quaternions (1843).

The generalised form of the scalar product, known as a symmetric bilinear form, has been of great importance in the twentieth-century study of linear groups.

Determinants were first used in Europe by G. W. Leibniz (1678) for the solution of simultaneous equations and the standard notation since the late eighteenth century has been that of qn 12.

Answers to chapter 14

1 $ab^2 = 0a^2 + 0b^2 - 2 \cdot 0a \cdot 0b \cos a0b$.

2 Either $\mathbf{a}$ or $\mathbf{b} = \mathbf{0}$ or angle $a0b = \frac{1}{2}\pi$.

3 Length of $0a = \sqrt{(\mathbf{a} \cdot \mathbf{a})}$.

4 $(x, y, z) \mapsto (x, y, z) \begin{pmatrix} a_1 \\ a_2 \\ a_3 \end{pmatrix}$. Codomain is $V_1(F) = F$. Kernel $=$ plane $\perp 0a$.

5

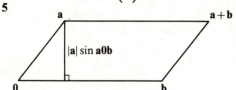

6 $(x, y, z) \mapsto (x, y, z) \begin{pmatrix} 0 & -a_3 & a_2 \\ a_3 & 0 & -a_1 \\ -a_2 & a_1 & 0 \end{pmatrix}$.

7 $\mathbf{a} \cdot (\mathbf{a} \times \mathbf{b}) = \mathbf{b} \cdot (\mathbf{a} \times \mathbf{b}) = 0$, so both $\mathbf{a}$ and $\mathbf{b}$ lie in the kernel of $\mathbf{v} \mapsto \mathbf{v} \cdot (\mathbf{a} \times \mathbf{b})$. Line joining origin to $\mathbf{a} \times \mathbf{b}$ is perpendicular to the lines joining the origin to $\mathbf{a}$ and to $\mathbf{b}$. $\mathbf{v} \mapsto \mathbf{v} \cdot (\mathbf{b} \times \mathbf{c})$ is linear by qn 4. So $(l\mathbf{a} + m\mathbf{b} + n\mathbf{c}) \cdot (\mathbf{b} \times \mathbf{c}) = l\mathbf{a} \cdot (\mathbf{b} \times \mathbf{c}) + m\mathbf{b} \cdot (\mathbf{b} \times \mathbf{c}) + n\mathbf{c} \cdot (\mathbf{b} \times \mathbf{c})$.

8 Composite of $\mathbf{v} \mapsto \mathbf{v} \times \mathbf{c}$ and $\mathbf{v} \mapsto \mathbf{v} \cdot \mathbf{a} = \mathbf{a} \cdot \mathbf{v}$. Codomain is $V_1(F) = F$. $\mathbf{a}$ lies in kernel from qn 7. $\mathbf{c}$ lies in kernel from qn 6.

9 Composite of $\mathbf{v} \mapsto \mathbf{b} \times \mathbf{v}$ and $\mathbf{v} \mapsto \mathbf{a} \cdot \mathbf{v} = \mathbf{v} \cdot \mathbf{a}$.

10 Use qn 5. If three leading edges are coplanar, the volume of the parallelepiped is 0.

11 All 0.

12 $\mathbf{a} \cdot (\mathbf{b} \times \mathbf{c}) = a_1(b_2 c_3 - b_3 c_2) + a_2(b_3 c_1 - b_1 c_3) + a_3(b_1 c_2 - b_2 c_1)$. Even permutations have a positive coefficient, odd permutations have a negative coefficient.

13 $\mathbf{a} \cdot \mathbf{b} \times \mathbf{c} = \mathbf{b} \cdot \mathbf{c} \times \mathbf{a} = \mathbf{c} \cdot \mathbf{a} \times \mathbf{b} = -\mathbf{a} \cdot \mathbf{c} \times \mathbf{b} = -\mathbf{b} \cdot \mathbf{c} \times \mathbf{a} = -\mathbf{c} \cdot \mathbf{b} \times \mathbf{a}$.

14 $(\mathbf{a} \cdot \mathbf{b} \times \mathbf{c}) I$.

15 (i) $\begin{pmatrix} 1 & -1 & 1 \\ -1 & 1 & 0 \\ 0 & 1 & -1 \end{pmatrix}$, (ii) $\begin{pmatrix} \frac{3}{2} & -\frac{1}{4} & -\frac{9}{4} \\ -1 & \frac{1}{2} & \frac{3}{2} \\ \frac{1}{2} & -\frac{1}{4} & -\frac{1}{4} \end{pmatrix}$,

(iii) $\dfrac{1}{16}\begin{pmatrix} 6 & 3 & -1 \\ -4 & 6 & -2 \\ 2 & 1 & 5 \end{pmatrix}$.

16 If $\mathbf{a} \cdot \mathbf{b} \times \mathbf{c} \neq 0$ then $A(1/(\mathbf{a} \cdot \mathbf{b} \times \mathbf{c}))B = I$.

17 Compare details with expansion in qn 12. If B is the right inverse of A^T, then $A^T B = I$ so $(A^T B)^T = I^T$ and $B^T A = I$.

22 If $AA^{-1} = I$ then $\det A \cdot \det A^{-1} = 1$ so $\det A \neq 0$.

23 Use qns 1.24 and 1.28.

24 Argue as in qns 13.13 and 13.14.

25 $(1, 0, 0)\alpha = \mathbf{a}$, $(0, 1, 0)\alpha = \mathbf{b}$, $(0, 0, 1)\alpha = \mathbf{c}$, so $(x, y, z)\alpha = x\mathbf{a} + y\mathbf{b} + z\mathbf{c}$. $\det A = 0 \Rightarrow \mathbf{a}$ in kernel. From qn 7, $\mathbf{b}$ and $\mathbf{c}$ are in the kernel. The kernel is the whole space only if $\mathbf{b} \times \mathbf{c} = \mathbf{0}$, but this only occurs if $\mathbf{b} \in Sp(\mathbf{c})$ or $\mathbf{c} \in Sp(\mathbf{b})$ from qn 6, and then we have $Sp(\mathbf{a}, \mathbf{b}, \mathbf{c}) = Sp(\mathbf{a}, \mathbf{b})$ or $Sp(\mathbf{a}, \mathbf{c})$.

26 A singular $\Rightarrow A^T$ singular. If A^T has rows $\mathbf{a}$, $\mathbf{b}$, $\mathbf{c}$ then B constructed as in qn 14 makes $A^T B$ = zero matrix. $B^T A = (A^T B)^T$ = zero matrix, so the rows of B^T, i.e. $\mathbf{b} \times \mathbf{c}$, $\mathbf{c} \times \mathbf{a}$ and $\mathbf{a} \times \mathbf{c}$ are mapped to $\mathbf{0}$. At least one of these is nonzero unless $\mathbf{a}$, $\mathbf{b}$, $\mathbf{c}$ are in a 1-dimensional subspace. If these lie in a 1-dimensional subspace then any vector perpendicular to it is in the kernel.

27 Enough to guarantee commutativity with

$$\begin{pmatrix} 1 & 0 & 1 \\ 0 & 1 & 0 \\ 0 & 0 & 1 \end{pmatrix}, \quad \begin{pmatrix} 1 & 1 & 0 \\ 0 & 1 & 0 \\ 0 & 0 & 1 \end{pmatrix}$$

and their transposes.

28 $p\mathbf{a} + q\mathbf{b} + r\mathbf{c} = 0$. Every point on $px + qy + rz = 0$ is fixed.

29 $1, 4, 8, 168 = 7 \cdot 6 \cdot 4$.

15

Eigenvectors and eigenvalues

The actual matrix for a linear transformation often hides the geometrical character of the transformation. For many linear transformations a geometrical view may emerge from finding those vectors which are transformed as if by an enlargement with centre the origin. These vectors are called the eigenvectors of the transformation and their related scale factors are called their eigenvalues.

1 Under the linear transformation of $V_2(\mathbb{R})$ given by

$$\alpha : (x, y) \mapsto (x, y)\begin{pmatrix} 1 & 0 \\ 0 & -1 \end{pmatrix},$$

which represents a reflection on the x-axis, determine whether $\mathbf{0}$, $\mathbf{v}$ and $\mathbf{v}\alpha$ are collinear points when (i) $\mathbf{v} = (1, 0)$, (ii) $\mathbf{v} = (0, 1)$ and (iii) $\mathbf{v} = (1, 1)$.

2 Under the linear transformation of $V_2(\mathbb{R})$ given by

$$\alpha : (x, y) \mapsto (x, y)\begin{pmatrix} 1 & 0 \\ 1 & 1 \end{pmatrix},$$

which represents a shear on the x-axis, determine whether $\mathbf{0}$, $\mathbf{v}$ and $\mathbf{v}\alpha$ are collinear points when (i) $\mathbf{v} = (1, 0)$, (ii) $\mathbf{v} = (0, 1)$ and (iii) $\mathbf{v} = (1, 1)$.

3 Under the linear transformation of $V_2(\mathbb{R})$ given by

$$\alpha : (x, y) \mapsto (x, y)\begin{pmatrix} 2 & 0 \\ 0 & 3 \end{pmatrix},$$

which is known as a *two-way stretch*, determine whether $\mathbf{0}$, $\mathbf{v}$ and $\mathbf{v}\alpha$ are collinear points when (i) $\mathbf{v} = (1, 0)$, (ii) $\mathbf{v} = (5, 0)$, (iii) $\mathbf{v} = (0, 1)$, (iv) $\mathbf{v} = (0, 4)$, (v) $\mathbf{v} = (1, 1)$.

4 If $\mathbf{0}$, $\mathbf{v}$ and $\mathbf{v}\alpha$ are collinear, how may $\mathbf{v}\alpha$ be expressed in terms of $\mathbf{v}$?

5 If $\alpha: V_n(F) \to V_n(F)$ is a linear transformation, then a vector $\mathbf{v} \neq \mathbf{0}$ such that $\mathbf{v}\alpha = \lambda\mathbf{v}$ for some scalar λ is called an *eigenvector* of α with *eigenvalue* λ.

What are the possible eigenvalues when α is a rotation of $V_2(\mathbb{R})$ about the origin?

6 If a linear transformation is an isometry, what real eigenvalues may it have?

7 Illustrate the image of a square lattice in $V_2(\mathbb{R})$ under the transformation

$$(x, y) \mapsto (x, y)\begin{pmatrix} 3 & 1 \\ 2 & 2 \end{pmatrix}$$

and then identify the images of the points $(2, 1)$ and $(1, -1)$. Are these eigenvectors of the transformation and if so, what are their eigenvalues?

8 If α is a linear transformation of $V_n(F)$ to itself, explain why every nonzero vector in the kernel of α is an eigenvector of α.

9 If $(1, 0)$ is an eigenvector of

$$\alpha: (x, y) \mapsto (x, y)\begin{pmatrix} a & b \\ c & d \end{pmatrix} \text{ what can be said about } b?$$

If $(0, 1)$ is an eigenvector of α, what can be said about c?
If $(1, 0)$, $(0, 1)$ and $(1, 1)$ are eigenvectors of α what can be said about a, b, c, d? How would you then describe the transformation α geometrically?
If $a = d$ and $b = c = 0$, what are the eigenvectors of α?

10 If $\mathbf{v}$ is an eigenvector of a linear transformation α, does it follow that $2\mathbf{v}$, $3\mathbf{v}$ and in fact $k\mathbf{v}$ are eigenvectors of α for every scalar k?

11 If $\mathbf{u}$ and $\mathbf{v}$ are eigenvectors of a linear transformation α and both $\mathbf{u}$ and $\mathbf{v}$ have the same eigenvalue λ, what conclusion may be drawn about the vectors of $Sp(\mathbf{u}, \mathbf{v})$?

Characteristic equation

12 If $\mathbf{u}$ is an eigenvector of the linear transformation

$$\mathbf{v} \mapsto \mathbf{v}\begin{pmatrix} a & b \\ c & d \end{pmatrix},$$

and $\mathbf{u}$ has eigenvalue λ, what can be said about $\mathbf{u}$ in relation to the linear transformation

$$\mathbf{v} \mapsto \mathbf{v}\begin{pmatrix} a - \lambda & b \\ c & d - \lambda \end{pmatrix}?$$

Deduce from qns 13.3, 13.4 and 13.5 that the matrix $\begin{pmatrix} a - \lambda & b \\ c & d - \lambda \end{pmatrix}$ is singular.

13 The equation $\det \begin{pmatrix} a - \lambda & b \\ c & d - \lambda \end{pmatrix} = 0$ is called the *characteristic equation* of the transformation $\mathbf{v} \mapsto \mathbf{v} \begin{pmatrix} a & b \\ c & d \end{pmatrix}$ and of the matrix $\begin{pmatrix} a & b \\ c & d \end{pmatrix}$.

By solving the characteristic equations of the transformations

(i) $(x, y) \mapsto (x, y) \begin{pmatrix} 3 & 1 \\ 2 & 2 \end{pmatrix}$,

(ii) $(x, y) \mapsto (x, y) \begin{pmatrix} 1 & 4 \\ 1 & 1 \end{pmatrix}$,

(iii) $(x, y) \mapsto (x, y) \begin{pmatrix} 4 & 2 \\ -1 & 1 \end{pmatrix}$,

(iv) $(x, y) \mapsto (x, y) \begin{pmatrix} 2 & 1 \\ -1 & 0 \end{pmatrix}$,

(v) $(x, y) \mapsto (x, y) \begin{pmatrix} 1 & 2 \\ 2 & 4 \end{pmatrix}$,

(vi) $(x, y) \mapsto (x, y) \begin{pmatrix} \cos\theta & \sin\theta \\ \sin\theta & -\cos\theta \end{pmatrix}$,

determine the possible eigenvalues of these transformations.

14 From the equation $(x, y) \begin{pmatrix} 3 & 1 \\ 2 & 2 \end{pmatrix} = (4x, 4y)$, determine the eigenvectors of qn 13(i) with eigenvalue 4. Similarly, determine the eigenvectors of this transformation with eigenvalue 1.

15 Find the eigenvectors of the transformation qn 13(iii).

16 Find the eigenvectors of the transformation

$$(x, y) \mapsto (x, y) \begin{pmatrix} a & 0 \\ 0 & b \end{pmatrix},$$

assuming $a \neq b$.

17 Let A denote the 3×3 matrix

$$\begin{pmatrix} a_1 & a_2 & a_3 \\ b_1 & b_2 & b_3 \\ c_1 & c_2 & c_3 \end{pmatrix}$$

Use an argument similar to that of qn 12 to show that if the transformation $\mathbf{v} \mapsto \mathbf{v}A$ has an eigenvector with eigenvalue λ, then the matrix

$$\begin{pmatrix} a_1 - \lambda & a_2 & a_3 \\ b_1 & b_2 - \lambda & b_3 \\ c_1 & c_2 & c_3 - \lambda \end{pmatrix}$$

is singular.

This matrix is often denoted by $A - \lambda I$ and the equation $\det (A - \lambda I) = 0$ is called the *characteristic equation* both of the transformation $\mathbf{v} \mapsto \mathbf{v}A$ and of the matrix A.

18 By first solving its characteristic equation, find the eigenvalues and eigenvectors of the transformation $\mathbf{v} \mapsto \mathbf{v}A$, where

$$A = \begin{pmatrix} 2 & 1 & 1 \\ -2 & 1 & 3 \\ 3 & 1 & -1 \end{pmatrix}.$$

Similar matrices

19 Let A be an $n \times n$ matrix, and let $\mathbf{u}$ be an eigenvector of the linear transformation $\mathbf{v} \mapsto \mathbf{v}A$, with eigenvalue λ, let U be a nonsingular $n \times n$ matrix and let $\mathbf{w} = \mathbf{u}U^{-1}$.

Find the images of $\mathbf{w}$ under the transformations

$$\mathbf{v} \mapsto \mathbf{v}U, \; \mathbf{v} \mapsto \mathbf{v}UA \text{ and } \mathbf{v} \mapsto \mathbf{v}UAU^{-1}$$

and deduce that $\mathbf{w}$ is an eigenvector of $\mathbf{v} \mapsto \mathbf{v}UAU^{-1}$ with eigenvalue λ. Illustrate this result on a drawing of $Sp(\mathbf{w})$ and $Sp(\mathbf{u})$.

20 Let $\mathbf{v} \mapsto \mathbf{v}A$ be a linear transformation of $V_2(F)$ with eigenvectors $\mathbf{u}_1$ and $\mathbf{u}_2$ with eigenvalues λ_1 and λ_2 respectively. Suppose further that the matrix U with rows $\mathbf{u}_1$ and $\mathbf{u}_2$ is nonsingular. Find the images of $(1, 0)$ and $(0, 1)$ under the transformation

$$\mathbf{v} \mapsto \mathbf{v}UAU^{-1},$$

and deduce that

$$UAU^{-1} = \begin{pmatrix} \lambda_1 & 0 \\ 0 & \lambda_2 \end{pmatrix}.$$

21 Use qn 15 to find a matrix U such that

$$U \begin{pmatrix} 4 & 2 \\ -1 & 1 \end{pmatrix} U^{-1} = \begin{pmatrix} 2 & 0 \\ 0 & 3 \end{pmatrix}.$$

Check your result.

22 Find a matrix U and a diagonal matrix D such that

$$U \begin{pmatrix} -5 & 2 \\ -28 & 10 \end{pmatrix} U^{-1} = D.$$

Show that there are two distinct possible diagonal matrices D for appropriate choices of U.

23 Let $\mathbf{v} \mapsto \mathbf{v}A$ be a linear transformation of $V_3(F)$ for which $\mathbf{u}_1$, $\mathbf{u}_2$ and $\mathbf{u}_3$ are eigenvectors with eigenvalues λ_1, λ_2 and λ_3 respectively.

Suppose further that the matrix U with rows $\mathbf{u}_1$, $\mathbf{u}_2$ and $\mathbf{u}_3$ is non-singular. Determine the matrix UAU^{-1}.

24 If $\mathbf{v} \mapsto \mathbf{v}A$ is a linear transformation of $V_3(F)$ for which $\mathbf{u}_1$, $\mathbf{u}_2$ and $\mathbf{u}_3$ are eigenvectors with eigenvalues λ_1, λ_2 and λ_3 respectively, find a matrix U such that

$$UA = \begin{pmatrix} \lambda_1 & 0 & 0 \\ 0 & \lambda_2 & 0 \\ 0 & 0 & \lambda_3 \end{pmatrix} U$$

whether or not U is nonsingular.

25 Is there more than one matrix U which satisfies the conditions of qn 24?

26 The matrices A and B are said to be *similar* if for some matrix M, $B = M^{-1}AM$.

Show that similarity is an equivalence relation on the set of $n \times n$ matrices over a given field.

Use qn 19 to show that similar matrices have the same eigenvalues.

27 If $\begin{pmatrix} a_1 & b_1 \\ c_1 & d_1 \end{pmatrix}$ and $\begin{pmatrix} a_2 & b_2 \\ c_2 & d_2 \end{pmatrix}$ are similar matrices, by considering their characteristic equations, prove that $a_1 + d_1 = a_2 + d_2$ and $a_1 d_1 - b_1 c_1 = a_2 d_2 - b_2 c_2$. Also give a direct proof that similar matrices have the same determinant.

Change of basis

28 For which pairs of vectors $\mathbf{u}$, $\mathbf{v}$ does $Sp(\mathbf{u}, \mathbf{v}) = V_2(F)$?

(i) $\mathbf{u} = (1, 0)$, $\mathbf{v} = (0, 1)$,

(ii) $\mathbf{u} = (1, 1)$, $\mathbf{v} = (0, 0)$,

(iii) $\mathbf{u} = (1, -2)$, $\mathbf{v} = (-2, 4)$,

(iv) $\mathbf{u} = (1, 2)$, $\mathbf{v} = (1, 3)$.

When $Sp(\mathbf{u}, \mathbf{v}) = V_2(F)$, the vectors $\mathbf{u}$ and $\mathbf{v}$ are said to form a *basis* for $V_2(F)$.

29 If α is a linear transformation of $V_n(F)$ with eigenvectors $\mathbf{u}$ and $\mathbf{v}$ with eigenvalues λ and μ, respectively, use qn 10 to show that if $\mathbf{v} \in Sp(\mathbf{u})$ or $\mathbf{u} \in Sp(\mathbf{v})$, then $\lambda = \mu$.

30 If α is a linear transformation of $V_2(F)$ with eigenvectors $\mathbf{u}$ and $\mathbf{v}$ with eigenvalues λ and μ, respectively, use qn 11.25 to show that if $\lambda \neq \mu$, then the matrix $\begin{pmatrix} \mathbf{u} \\ \mathbf{v} \end{pmatrix}$ is nonsingular and deduce that $\mathbf{u}$ and $\mathbf{v}$ form a basis for $V_2(F)$.

31 If the linear transformation $\mathbf{v} \mapsto \mathbf{v}A$ of $V_2(F)$ has two distinct eigen-values, deduce from qn 20 that A is similar to a diagonal matrix.

32 If α is a linear transformation of $V_n(F)$ with eigenvectors $\mathbf{u}$, $\mathbf{v}$ and $\mathbf{w}$ with eigenvalues λ, μ and ν, respectively, and $\mathbf{w} \in Sp(\mathbf{u}, \mathbf{v})$, use qn 11.28 to show that either $\nu = \lambda$, $\nu = \mu$ or $\lambda = \mu$.

33 If a 2×2 matrix A has just one eigenvalue λ, is it possible that there exists a basis of $V_2(F)$ consisting of eigenvectors of $\mathbf{v} \mapsto \mathbf{v}A$? Is there necessarily a basis of $V_2(F)$ consisting of eigenvectors of $\mathbf{v} \mapsto \mathbf{v}A$?

34 If $Sp(\mathbf{u}, \mathbf{w}) = V_2(F)$, then every vector $\mathbf{v}$ in $V_2(F)$ has the form $x\mathbf{u} + y\mathbf{w}$, and for a given $\mathbf{v}$, the scalars x and y are unique from qn 11.28. When $\mathbf{v} = (2, 3)$, find x and y when

(i) $\mathbf{u} = (1, 0)$ and $\mathbf{w} = (0, 1)$,

(ii) $\mathbf{u} = (1, 2)$ and $\mathbf{w} = (1, 3)$,

(iii) $\mathbf{u} = (1, 1)$ and $\mathbf{w} = (1, -1)$.

The scalars x and y are called the coordinates of $\mathbf{v}$ relative to the basis $\mathbf{u}$, $\mathbf{w}$.

35 If $\alpha : \mathbf{v} \mapsto \mathbf{v}A$ is a linear transformation of $V_2(F) = Sp(\mathbf{u}, \mathbf{w})$, and $\mathbf{u}A = p\mathbf{u} + q\mathbf{w}$ and $\mathbf{w}A = r\mathbf{u} + s\mathbf{w}$, deduce that

$$\begin{pmatrix} \mathbf{u} \\ \mathbf{w} \end{pmatrix}A = \begin{pmatrix} p & q \\ r & s \end{pmatrix}\begin{pmatrix} \mathbf{u} \\ \mathbf{w} \end{pmatrix},$$

so that

$$\begin{pmatrix} p & q \\ r & s \end{pmatrix} = \begin{pmatrix} \mathbf{u} \\ \mathbf{w} \end{pmatrix}A\begin{pmatrix} \mathbf{u} \\ \mathbf{w} \end{pmatrix}^{-1}.$$

If, further, we denote the vector $x\mathbf{u} + y\mathbf{w}$ by $[x, y]$, show that $\alpha : [x, y] \mapsto [x, y]\begin{pmatrix} p & q \\ r & s \end{pmatrix}$, so that similar matrices represent the same transformation with respect to different bases.

Shears

36 Use the definition of a shear in qn 13.23 to show that a shear has eigenvectors with eigenvalue 1 and no other eigenvalues.

37 What is the characteristic equation for the matrix of a shear of $V_2(F)$?

38 If $\mathbf{v} \mapsto \mathbf{v}A$ is a shear of $V_2(F)$, what is the determinant of A? What is the effect of a shear on areas of $V_2(\mathbb{R})$?

39 Express the linear transformation

$$(x, y) \mapsto (x, y) + (cx + dy)(a, b)$$

using a matrix. Check that the algebraic condition which is
necessary for this to be a shear implies that the characteristic
equation of this matrix is $(\lambda - 1)^2 = 0$.

40 If A is a 2×2 matrix with characteristic equation $(\lambda - 1)^2 = 0$, show
that A may be expressed in the form $\begin{pmatrix} 1 + a & b \\ c & 1 - a \end{pmatrix}$ where
$-a^2 - bc = 0$.

(i) If $b = 0$, show that this matrix is that of a shear as in qn 13.24,
or of the identity.
(ii) If $b \neq 0$, show that $\mathbf{v} \mapsto \mathbf{v}A$ has the form

$$(x, y) \mapsto (x, y) + \left(\frac{bx - ay}{b}\right)(a, b).$$

Questions 39 and 40 establish that the matrices for shears of $V_2(F)$
are precisely those different from the identity which have characteristic
equation $(\lambda - 1)^2 = 0$.

41 If a and b are given with $b \neq 0$, show that there is a unique matrix of
the form $\begin{pmatrix} a & b \\ r & s \end{pmatrix}$ which represents a shear.

42 If c and d are given with $c \neq 0$, show that there is a unique matrix of
the form $\begin{pmatrix} p & q \\ c & d \end{pmatrix}$ which represents a shear.

43 Use qn 13.9 to show that the subset of matrices with determinant 1 in
$GL(2, F)$ is a subgroup. This subgroup is called the *special linear
group SL(2, F)*.

44 List the elements of $SL(2, \mathbb{Z}_2)$ and say which are shears and which are
not. Do the shears generate the group?

45 If $A = \begin{pmatrix} a & b \\ c & d \end{pmatrix}$ and $S = \begin{pmatrix} a & b \\ r & s \end{pmatrix}$ are both matrices in $SL(2, F)$
show that the matrix AS^{-1} is that of a shear or of the identity. Use
qn 41 to show that if $b \neq 0$, then A is either the matrix of a shear
or of the product of two shears.

46 Use the ideas of qn 45 and the result of qn 42 to show that if
$A = \begin{pmatrix} a & b \\ c & d \end{pmatrix}$ is in $SL(2, F)$ and $c \neq 0$, then A is either the matrix
of a shear or of the product of two shears.

47 If $\alpha : (x, y) \mapsto (x, y)\begin{pmatrix} a & 0 \\ 0 & 1/a \end{pmatrix}$ and $a \neq 0, -1$, show that α is the
product of the two shears

$$(x, y) \mapsto (x, y) + \frac{a-1}{a+1}(x - y)(1, 1), \text{ and}$$

$$(x, y) \mapsto (x, y) + \frac{a-1}{a+1}(x + ay)(1, -1/a).$$

Questions 45, 46 and 47 conclude the proof that every element of $SL(2, F)$ is either a shear, the product of two shears or the half-turn $(x, y) \mapsto (-x, -y)$.

Summary

Definition If $\alpha: V \to V$ is a linear transformation of a vector space
qn 5 V, and for some vector $\mathbf{u} \neq \mathbf{0}$, $\mathbf{u}\alpha = \lambda\mathbf{u}$ for some scalar
λ, then $\mathbf{u}$ is said to be an *eigenvector* of α with *eigenvalue* λ.

Definition The equation det $(A - \lambda I) = 0$ is called the *charac-*
qns 13, 17 *teristic equation* of the matrix A and of the transformation $\mathbf{v} \mapsto \mathbf{v}A$.

Theorem The linear transformation $\mathbf{v} \mapsto \mathbf{v}A$ has an eigenvector
qns 12, 17 with eigenvalue λ if and only if det $(A - \lambda I) = 0$.

Theorem If $\mathbf{u}$ is an eigenvector of the linear transformation
qn 19 $\mathbf{v} \mapsto \mathbf{v}A$ with eigenvalue λ, and M is a nonsingular
matrix, then $\mathbf{u}M^{-1}$ is an eigenvector of the transformation $\mathbf{v} \mapsto \mathbf{v}MAM^{-1}$ with eigenvalue λ.

Theorem If the rows of the nonsingular matrix M are eigenvec-
qns 20, 23 tors of the transformation $\mathbf{v} \mapsto \mathbf{v}A$, then MAM^{-1} is a
diagonal matrix, and the diagonal entries in MAM^{-1}
are the eigenvalues of $\mathbf{v} \mapsto \mathbf{v}A$.

Definition Two matrices A and B are said to be *similar* when there
qn 26   exists a matrix M such that $B = M^{-1}AM$.

Theorem Similar matrices have the same eigenvalues, the same
qns 26, 27 characteristic equation and the same determinant.

Theorem Similar matrices represent the same transformation with
qn 35   respect to different bases.

Theorem The only transformations in $GL(2, F)$ with character-
qns 39, 40 istic equation $(\lambda - 1)^2 = 0$ are the shears and the identity.

Definition The subgroup of $GL(2, F)$ of matrices with deter-
qn 43   minant 1 is called the *special linear group* $SL(2, F)$.

Theorem The special linear group is generated by shears.
qns 45,
46, 47

Historical note

The use of eigenvectors and eigenvalues first appeared in the study of quadratic forms. The problem was to find a rotation which would transform the curve $ax^2 + 2hxy + by^2 = c$ into a curve of the form $Ax^2 + By^2 = C$. The method is to write $ax^2 + 2hxy + by^2$ in the

form $(x, y)\begin{pmatrix} a & h \\ h & b \end{pmatrix}\begin{pmatrix} x \\ y \end{pmatrix}$ and then A and B are the eigenvalues of the

matrix $\begin{pmatrix} a & h \\ h & b \end{pmatrix}$. If $\mathbf{a}$ and $\mathbf{b}$ are corresponding eigenvectors of unit

length, then the matrix with rows $\mathbf{a}$ and $\mathbf{b}$ is that of the required rotation. In fact it was in attempting to perform the appropriate rotation for surfaces in three and higher dimensions that these concepts and methods were developed by A. Cayley and others during the 1840s. These methods appear, though not in modern language, in the text of G. Salmon (1859).

   A. Cayley's memoir on matrices (1858) includes the result that a 2×2 matrix satisfies its own characteristic equation.

Answers to chapter 15

1 (i) yes, (ii) yes, (iii) no.

2 (i) yes, (ii) no, (iii) no.

3 (i) yes, (ii) yes, (iii) yes, (iv) yes, (v) no.

4 $\mathbf{v}\alpha$ = a scalar multiple of $\mathbf{v}$.

5 Only ± 1.

6 Only ± 1.

7 $(2, 1) \mapsto (8, 4)$, $(1, -1) \mapsto (1, -1)$, eigenvalues 4, 1.

8 Eigenvalue 0.

9 $b = 0$. $c = 0$. With all three eigenvectors, we have $a = d$ as well, then α is an enlargement, and every vector is an eigenvector.

10 $\mathbf{v}\alpha = \lambda\mathbf{v} \Rightarrow (k\mathbf{v})\alpha = k(\mathbf{v}\alpha) = k(\lambda\mathbf{v}) = \lambda(k\mathbf{v})$.

11 $\mathbf{u}\alpha = \lambda\mathbf{u}$ and $\mathbf{v}\alpha = \lambda\mathbf{v} \Rightarrow (a\mathbf{u} + b\mathbf{v})\alpha = a(\lambda\mathbf{u}) + b(\lambda\mathbf{v}) = \lambda(a\mathbf{u} + b\mathbf{v})$.

12 $\mathbf{u}$ is the kernel, since $\mathbf{u}\begin{pmatrix} a & b \\ c & d \end{pmatrix} = \mathbf{u}\begin{pmatrix} \lambda & 0 \\ 0 & \lambda \end{pmatrix}$.

13 (i) 4, 1, (ii) 3, -1, (iii) 3, 2, (iv) 1, (v) 5, 0, (vi) ± 1.

14 Multiples of $(2, 1)$ have eigenvalue 4, multiples of $(-1, 1)$ have eigenvalue 1.

15 Multiples of $(1, 1)$ have eigenvalue 3, multiples of $(1, 2)$ have eigenvalue 2.

16 Multiples of $(1, 0)$ have eigenvalue a, multiples of $(0, 1)$ have eigenvalue b.

18 $(-1, 1, 1)$ has eigenvalue 1. $(11, 1, -14)$ has eigenvalue -2. $(1, 1, 1)$ has eigenvalue 3.

19 $(\mathbf{u}U^{-1})U = \mathbf{u}$, $(\mathbf{u}U^{-1})UA = \mathbf{u}A = \lambda\mathbf{u}$, $(\mathbf{u}U^{-1})UAU^{-1} = \lambda\mathbf{u}U^{-1}$. $\mathbf{w}UAU^{-1} = \lambda\mathbf{w}$.

20 $(1, 0)UAU^{-1} = \mathbf{u}_1 AU^{-1} = \lambda_1\mathbf{u}_1 U^{-1} = \lambda_1(1, 0)$.

21 $U = \begin{pmatrix} 1 & 2 \\ 1 & 1 \end{pmatrix}$ is the natural choice. $U = \begin{pmatrix} a & 2a \\ b & b \end{pmatrix}$ will also work.

22 $U = \begin{pmatrix} -4 & 1 \\ 7 & -2 \end{pmatrix}$ for $D = \begin{pmatrix} 2 & 0 \\ 0 & 3 \end{pmatrix}$. $U = \begin{pmatrix} 7 & -2 \\ -4 & 1 \end{pmatrix}$ for $D = \begin{pmatrix} 3 & 0 \\ 0 & 2 \end{pmatrix}$.

23 $\begin{pmatrix} \lambda_1 & 0 & 0 \\ 0 & \lambda_2 & 0 \\ 0 & 0 & \lambda_3 \end{pmatrix}$.

24, 25 U has rows $\mathbf{u}_1, \mathbf{u}_2$ and $\mathbf{u}_3$ or scalar multiples of these.

26 Reflexive, for $M = I$. Symmetric for $B = M^{-1}AM \Rightarrow A = (M^{-1})^{-1}BM^{-1}$. Transitive, for $B = M^{-1}AM$, $C = N^{-1}BN \Rightarrow C = (MN)^{-1}A(MN)$.

27 From qn 26, the characteristic equations have the same roots and since the coefficients of λ^2 are the same, the other corresponding coefficients must be equal. A general proof that similar matrices have the same characteristic equation (irrespective of the existence of roots) is indicated in qn 19.11. $\det (M^{-1}AM) = \det M^{-1} \cdot \det A \cdot \det M = \det A \cdot \det M^{-1} \cdot \det M = \det A \cdot \det M^{-1}M$.

28 (i) and (iv).

30 If $\begin{pmatrix} \mathbf{u} \\ \mathbf{v} \end{pmatrix}$ is singular, $Sp(\mathbf{u}, \mathbf{v})$ is 1-dimensional allowing at most one eigenvalue.

31 Two eigenvectors form a basis. Use these as the rows of U.

32 If $\mathbf{w} = a\mathbf{u} + b\mathbf{v}$, then $a = 0 \Rightarrow v = \mu$ and $b = 0 \Rightarrow v = \lambda$. Suppose $a \neq 0 \neq b$, then $\mathbf{w}\alpha = (a\mathbf{u} + b\mathbf{v})\alpha$ and $v(a\mathbf{u} + b\mathbf{v}) = a\lambda\mathbf{u} + b\mu\mathbf{v}$, so $a(\lambda - v) = 0$ and $b(\mu - v) = 0$ and $\lambda = \mu = v$.

33 If $A = \begin{pmatrix} \lambda & 0 \\ 0 & \lambda \end{pmatrix}$, any two basis vectors are eigenvectors.

If $A = \begin{pmatrix} \lambda & 0 \\ 1 & \lambda \end{pmatrix}$, all eigenvectors have the form $(a, 0)$.

34 (i) $[x, y] = [2, 3]$, (ii) $[x, y] = [3, -1]$, (iii) $[x, y] = [\frac{5}{2}, -\frac{1}{2}]$.

35 $\mathbf{u} = [1, 0]$. $\alpha : [1, 0] \mapsto [p, q]$. $\mathbf{w} = [0, 1]$. $\alpha : [0, 1] \mapsto [r, s]$.

36 Obviously $\mathbf{a}$ has eigenvalue 1. If $\mathbf{v}$ has eigenvalue $\lambda \neq 1$ then $\lambda\mathbf{v} = \mathbf{v} + f(\mathbf{v})\mathbf{a}$ so $(\lambda - 1)\mathbf{v} = f(\mathbf{v})\mathbf{a}$ so $\mathbf{v} \in Sp(\mathbf{a})$. Contradiction.

37 $(\lambda - 1)^2 = 0$, from qn 36.

38 $\det A = 1$, from qn 37. Shears preserve areas.

39 $\begin{pmatrix} 1 + ac & bc \\ ad & 1 + bd \end{pmatrix}$. See qn 13.23.

40 If $\begin{pmatrix} a & b \\ c & d \end{pmatrix}$ has characteristic equation $(\lambda - 1)^2 = 0$ then $a + d = 2$ and $ad - bc = 1$.

41 $a + s = 2$, $as - br = 1 \Rightarrow s = 2 - a$, $r = (as - 1)/b$.

42 $p + d = 2$, $pd - qc = 1 \Rightarrow p = 2 - d$, $q = (pd - 1)/c$.

44 Identity $\begin{pmatrix} 1 & 0 \\ 0 & 1 \end{pmatrix}$; shears $\begin{pmatrix} 1 & 0 \\ 1 & 1 \end{pmatrix}, \begin{pmatrix} 1 & 1 \\ 0 & 1 \end{pmatrix}, \begin{pmatrix} 0 & 1 \\ 1 & 0 \end{pmatrix}$; others $\begin{pmatrix} 0 & 1 \\ 1 & 1 \end{pmatrix}, \begin{pmatrix} 1 & 1 \\ 1 & 0 \end{pmatrix}$. Shears generate the group, like transpositions in S_3.

45 $(1, 0)AS^{-1} = (1, 0)$, so AS^{-1} has an eigenvalue 1. But det $AS^{-1} = 1$, so product of eigenvalues is 1 and so characteristic equation is $(\lambda - 1)^2$, so AS^{-1} is a shear or the identity. From qn 41, let S be the unique shear with first row (a, b). Then $AS^{-1} = T$ (a shear or the identity). If T is a shear then $A = ST$, the product of two shears.

16

Homomorphisms

Some groups exhibit, in miniature, the structure of other groups. A group H contains a 'miniature version' of a group G when there is a function of G to H, usually a many-to-one function, which preserves the group structure of G. Such a function is called a homomorphism. What we mean when we say that a function 'preserves the group structure' is that if $g_1 \mapsto h_1$ and $g_2 \mapsto h_2$, then $g_1 g_2 \mapsto h_1 h_2$. In ordinary language we describe this by saying that the image of a product is the product of the images.

Concurrent reading: Fraleigh, sections 11 and 12; Green, chapter 7.

1 If A and B are matrices in $GL(2, F)$ and A and B belong to the same right coset of $SL(2, F)$, what can you say about det A and det B?

2 If A and B are matrices in $GL(2, F)$ and det $A = $ det B, does it follow that A and B belong to the same right coset of $SL(2, F)$?

3 What is the index of $SL(2, \mathbb{Z}_3)$ in $GL(2, \mathbb{Z}_3)$ and what is the order of $SL(2, \mathbb{Z}_3)$? (See qn 13.32)

4 What is the index of $SL(2, \mathbb{Z}_5)$ in $GL(2, \mathbb{Z}_5)$? What is the order of $SL(2, \mathbb{Z}_5)$?

5 If A and B are matrices in $GL(2, F)$ and det $A = $ det B, does it follow that A and B belong to the same left coset of $SL(2, F)$?

6 Is $SL(2, F)$ a normal subgroup of $GL(2, F)$?

7 If $GL(2, F)$ is the domain of the function

$A \mapsto$ det A,

what is the set of images?

8 If a function $\alpha : G \to H$ maps a group $(G, \cdot)$ into a group $(H, \circ)$ and

$(g_1 \cdot g_2)\alpha = (g_1\alpha) \circ (g_2\alpha)$ for all $g_1, g_2 \in G$, then α is called a *group homomorphism*.

If $A \mapsto \det A$ is a group homomorphism of $GL(2, F)$, identify the operation in the image group.

9 The real number $|a|$ is called the *scale factor* of the similarity $z \mapsto az + b$. Is the function which carries each direct similarity into its scale factor a group homomorphism of the group of direct similarities onto the multiplicative group of positive real numbers?

10 Use arguments like those of qns 6.66 and 6.67 to show that if $\alpha: G \to H$ is a group homomorphism then the identity in G is mapped to the identity in H and pairs of elements which are inverse in G are mapped to pairs of elements which are inverse in H.

11 If $\alpha: G \to H$ is a group homomorphism, prove that the image of G is a subgroup of H. Prove also that the image of any subgroup of G is also a subgroup of H.

12 Is every linear transformation of a vector space a group homomorphism?

The kernel of a homomorphism

13 When the determinant is used to construct a group homomorphism of $GL(2, F)$ what is the subset of the domain whose image is the identity of the codomain?

14 When the scale factor is used to construct a group homomorphism of the group of direct similarities, what is the subset of the domain whose image is the identity of the codomain?

15 If $\alpha: G \to H$ is a group homomorphism, the full subset of G, each element of which has as image the identity in H is called the *kernel* of α. Prove that the kernel of α is a subgroup of G.

16 If K is the kernel of the group homomorphism $\alpha: G \to H$ and $g \in G$, show that every element of the coset Kg and every element of the coset gK has the same image in H.

17 If $g_1, g_2 \in G$ have the same image in H under the group homomorphism $\alpha: G \to H$, prove that they belong to the same left coset of the kernel of α and also that they belong to the same right coset of the kernel of α.

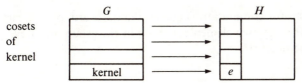

18 Prove that the kernel of a group homomorphism $\alpha : G \rightarrow H$ forms a normal subgroup of G.

19 If the kernel of a group homomorphism $\alpha : G \rightarrow H$ consists only of the identity in G, what can be said about the cosets of the kernel and hence about the function α?

Quotient groups

20 If N is a normal subgroup of a group G, prove that the cosets of N form a group under the operation of multiplication of subsets as in qn 9.32. This group is called the *quotient group* G/N.

21 From qn 9.31, exhibit the Cayley table of the group $D_4/\{e, a^2\}$.

22 For any group G with a normal subgroup N prove that the function $\alpha : G \rightarrow G/N$ given by $\alpha : g \mapsto Ng$ is a group homomorphism onto G/N.

23 *Fundamental theorem on group homomorphisms.* If $\alpha : G \rightarrow H$ is a group homomorphism with kernel K, show that the function from the image of G under α to the quotient group G/K given by $g\alpha \mapsto Kg$ is

  (i) well-defined,
  (ii) one–one,
  (iii) onto,
  (iv) structure-preserving, in the sense that

$$g_1\alpha \cdot g_2\alpha \mapsto Kg_1 \cdot Kg_2,$$

and thus establishes that every homomorphic image of a group is isomorphic to a quotient group.

The fundamental theorem implies that the problem of finding all homomorphic images of a group can be reduced to that of finding all normal subgroups and then constructing the corresponding quotient groups.

24 If G is a group with a subgroup N, and N has exactly two cosets in G, why must N be a normal subgroup? To what group is the quotient group G/N isomorphic?

25 Give examples of subgroups of index 2 in (i) S_n, (ii) D_n, (iii) the group of similarities of the Euclidean plane, (iv) the group of isometries of the Euclidean plane.

26 In an abelian (or commutative) group, that is, a group for which $ab = ba$ for every two elements a and b in the group, must every subgroup be normal?

The field $\mathbb{Z}_p$

27 If g is an element of a group $(G, \cdot)$ and a mapping $\alpha : \mathbb{Z} \to G$ is defined by

$\alpha : r \mapsto g^r$,

is α a group homomorphism of $(\mathbb{Z}, +)$?

28 If, in qn 27, the element g has order 3, list the elements of the kernel of α.

29 If in qn 27 the element g has order n, what can you say about g^a and g^b if $a \equiv b \pmod{n}$ with the notation of qn 9.2.

30 In qn 6.26 we found that all subgroups of $(\mathbb{Z}, +)$ were cyclic. What condition on the elements of G would allow you to construct a group homomorphism $\mathbb{Z} \to G$ with kernel N, for any subgroup N of $\mathbb{Z}$, chosen in advance?

31 The subgroup of $(\mathbb{Z}, +)$ generated by n consists of all multiples of n, and if we denote this group by $n\mathbb{Z}$, the quotient group $\mathbb{Z}/n\mathbb{Z}$ is conventionally denoted by $\mathbb{Z}_n$.
What are the elements of $\mathbb{Z}_n$ and how many of them are there? It is conventional to name the cosets of $n\mathbb{Z}$ with their smallest non-negative member. Find a generator for $\mathbb{Z}/n\mathbb{Z}$.

32 Since $(kn + a)(ln + b) = (kln + al + bk)n + ab$, the usual multiplication on the integers induces a consistent multiplication on the cosets of $n\mathbb{Z}$, so that $(\mathbb{Z}_n, +, \cdot)$ has the following properties:

$(\mathbb{Z}_n, +)$ is an abelian group,

$(\mathbb{Z}_n, \cdot)$ is closed, associative, commutative and has an identity.

$a \cdot (b + c) = a \cdot b + a \cdot c$ and $(a + b) \cdot c = a \cdot c + b \cdot c$.

Use qn 11.7 to prove that if n is a composite number, then $\mathbb{Z}_n$ is not a field with these operations.

33 Imagine the construction of a Cayley table for $(\mathbb{Z}_p - \{0\}, \cdot)$ where p is a prime number. Why is the set closed under multiplication? Argue by contradiction. Why are all the entries in one row of the Cayley table, entries like ax, ay, az, necessarily different? Again, argue by contradiction. How many different entries are there? Must 1 be one of the entries? If $ax \equiv 1 \pmod{p}$ what is the inverse of a in $(\mathbb{Z}_p - \{0\}, \cdot)$? Deduce that $(\mathbb{Z}_p - \{0\}, \cdot)$ is a group, and hence that $(\mathbb{Z}_p, +, \cdot)$ is a field.

34 By finding the order of the group $(\mathbb{Z}_p - \{0\}, \cdot)$ and considering the possible orders of elements in this group, prove that if x is not a

multiple of p, then $x^{p-1} \equiv 1 \pmod{p}$, and deduce that for all
integers x, $x^p \equiv x \pmod{p}$.

Two special quotient groups

35 Examine the additive group $\mathbb{Q}/\mathbb{Z}$, where $\mathbb{Q}$ denotes the set of rational
numbers. Show that it is an infinite group in which every element
has finite order.

36 If D is the group of direct isometries of the Euclidean plane, and T is
the translation group, prove that D/T is isomorphic to the multiplicative group of complex numbers with modulus 1.

Summary

Definition If $(G, \cdot)$ and $(H, \circ)$ are groups and $\alpha: G \to H$ is a func-
qn 8 tion such that $(g_1\alpha) \circ (g_2\alpha) = (g_1 \cdot g_2)\alpha$ for all $g_1, g_2 \in G$,
then α is said to be a *group homomorphism*.

Theorem Under a group homomorphism, the image of the iden-
qn 10, 11 tity is the identity, the images of an element and its
inverse are an element and its inverse, and the image of
a subgroup is a subgroup.

Definition The full subset of elements of the domain whose image
qn 15 is the identity under a group homomorphism is called
the *kernel* of the homomorphism.

Theorem The kernel of a group homomorphism is a normal sub-
qn 18 group of the domain.

Theorem If G is a group and N is a normal subgroup of G then
qn 20 the cosets of N form a group under multiplication of
subsets, called the *quotient group* G/N.

Funda- If G is a group then the image of G under a homomor-
mental phism α is isomorphic to the quotient group G/K, where
theorem K is the kernel of α.
qn 23

Historical note

Quotient groups have had a place in mathematical literature since
the time of C. F. Gauss' *Disquisitiones Arithmeticae* (1801) with its
modular arithmetic. E. Galois made use of normal subgroups and quotient groups in 1830. The notation G/N was introduced by C. Jordan
(1873). The explicit study of homomorphisms and their kernels began
with A. Capelli (1878).

Answers to chapter 16

1 $A = SB$ where $\det S = 1$ and $\det A = \det S \cdot \det B = \det B$.

2 $\det A = \det B \Rightarrow \det AB^{-1} = 1 \Rightarrow AB^{-1} = S$ where $\det S = 1$.

3 Number of cosets = number of distinct values of determinant = 2.
$|SL(2, \mathbb{Z}_3)| = |GL(2, \mathbb{Z}_3)|/2 = 24$.

4 4. $24 \cdot 20/4 = 120$.

5 $\det A = \det B \Rightarrow \det A^{-1}B = 1$. Yes.

6 Left and right cosets are subsets with same determinant. Yes.

7 $F - \{0\}$.

8 Multiplication.

9 If $\alpha : z \mapsto az + b$, $\gamma : z \mapsto cz + d$, $\alpha\gamma : z \mapsto caz + cb + d$. Now the function $\alpha \mapsto |a|$ takes $\alpha\gamma \mapsto |ca| = |a| \cdot |c|$. Yes.

11 Definition of homomorphism gives closure. Identity and inverses from qn 10.

12 Yes, of the group of vectors under vector addition.

13 $SL(2, F)$.

14 The subgroup of direct isometries.

15 $g_1\alpha = e\alpha$, $g_2\alpha = e\alpha \Rightarrow g_1\alpha \cdot g_2\alpha = e\alpha \cdot e\alpha \Rightarrow (g_1g_2)\alpha = e\alpha$, etc.

16 $(k_1g)\alpha = k_1\alpha \cdot g\alpha = g\alpha$. $(gk_2)\alpha = g\alpha \cdot k_2\alpha = g\alpha$.

17 $g_1\alpha = g_2\alpha \Rightarrow (g_1\alpha)(g_2\alpha)^{-1} = e\alpha \Rightarrow (g_1g_2^{-1})\alpha = e\alpha \Rightarrow g_1g_2^{-1} \in$ kernel. Also $(g_1^{-1}g_2)\alpha = e\alpha \Rightarrow g_1^{-1}g_2 \in$ kernel.

18 g_1, g_2 in same right coset $\Leftrightarrow g_1g_2^{-1} \in$ kernel $\Leftrightarrow g_1\alpha = g_2\alpha \Leftrightarrow g_1^{-1}g_2 \in$ kernel $\Leftrightarrow g_1, g_2$ in same left coset. Every left coset of kernel is a right coset.

19 Each coset is a singleton and the homomorphism is an injection (one–one). The homomorphism is then called an isomorphism.

20 Definition gives closure. Identity is N. Na^{-1} is inverse of Na.

21

$\{e, a^2\}$	$\{a, a^3\}$	$\{b, ba^2\}$	$\{ba, ba^3\}$
$\{a, a^3\}$	$\{e, a^2\}$	$\{ba, ba^3\}$	$\{b, ba^2\}$
$\{b, ba^2\}$	$\{ba, ba^3\}$	$\{e, a^2\}$	$\{a, a^3\}$
$\{ba, ba^3\}$	$\{b, ba^2\}$	$\{a, a^3\}$	$\{e, a^2\}$

22 Follows directly from qn 9.32.

23 (i) and (ii) $g_1\alpha = g_2\alpha \Leftrightarrow (g_1g_2^{-1})\alpha = e\alpha \Leftrightarrow g_1g_2^{-1} \in K \Leftrightarrow Kg_1 = Kg_2$. (iii) Obvious. (iv) $g_1\alpha \cdot g_2\alpha = (g_1g_2)\alpha \mapsto Kg_1g_2 = Kg_1 \cdot Kg_2$.

24 The cosets of N are N and $G - N$ which must be both left and right. Quotient group $\cong C_2$.

25 (i) A_n, (ii) C_n, (iii) direct similarities, (iv) direct isometries.

26 Yes.

27 $g^{m+n} = g^m \cdot g^n$. Yes.

28 $\{0, \pm 3, \pm 6, \ldots\}$.

29 $g^a = g^b$.

30 If $N = \langle n \rangle$ we need an element $g \in G$ of order n. Then map $1 \mapsto g$.

31 The elements of $\mathbb{Z}_n$ are the n cosets of $\langle n \rangle$ in $\mathbb{Z}$. $\langle n \rangle + 1$ generates $\mathbb{Z}_n$.

32 If $n = pq$ then $(kn + p)(ln + q)$ is a multiple of n and there are nonzero elements of $\mathbb{Z}_n$ whose product is zero.

33 Not closed only if $ax \equiv 0 \pmod{p}$. But a prime number has no proper divisors. $ax \equiv ay \Leftrightarrow a(x - y) \equiv 0 \Leftrightarrow x \equiv y \pmod{p}$. $p - 1$ different numbers for $p - 1$ different entries so all are there including 1. Argument establishes closure and existence of inverses in $(\mathbb{Z}_p - \{0\}, \cdot)$.

34 Group has order $p - 1$. So for any element x, we have $x^k \equiv 1$ with k a factor of $p - 1$. Thus $x^{p-1} \equiv 1 \pmod{p}$. When x is a multiple of p obviously both x and $x^p \equiv 0 \pmod{p}$ so $x^p \equiv x \pmod{p}$.

35 $\mathbb{Z} + p/q$ has order q.

36 If $\alpha : z \mapsto e^{i\theta}z + c$, the mapping $\alpha \mapsto e^{i\theta}$ establishes the isomorphism using qn 23.

17

Conjugacy

Every equivalence relation is a way of isolating a particular kind of sameness. The sameness which the equivalence relation of conjugacy identifies in the Euclidean group is the sameness between two reflections or the sameness between two half-turns. In the symmetric group, the sameness of conjugacy is the sameness of permutations with the same cycle structure. Much of the significance of this relation derives from the fact that the subgroups which are formed by the union of whole conjugacy classes are precisely the normal subgroups.

1 Express the permutations (1432) (12) (1234) and (132) (12) (123) as single cycles.

2 If α is an element of S_4, find the images of 1α, 2α, 3α and 4α under the permutation $\alpha^{-1}(12)\,\alpha$. Write the permutation $\alpha^{-1}(12)\,\alpha$ in cycle form.
The important point to remember is that $\alpha^{-1}(12)\,\alpha$ is always a transposition.

3 Exhibit the image of the set
$\{(1), (123), (132), (23), (13), (12)\}$
under the mapping given by
$\gamma \mapsto (1432)\gamma(1234)$.
Describe the original set and the image set as subsets of S_4.

Fixed points of conjugate elements

4 If α and β are permutations of a set containing the element m, prove that $m\alpha = m$ implies $(m\beta)\,\beta^{-1}\alpha\beta = m\beta$ and conversely.

5 If α is a rotation of the plane and β is any isometry of the plane, use qn 4 to show that $\beta^{-1}\alpha\beta$ has one and only one fixed point. What sort of isometry is $\beta^{-1}\alpha\beta$?

6 If α is a reflection of the plane and β is any isometry, what can you say about the fixed points of $\beta^{-1}\alpha\beta$? What kind of isometry is $\beta^{-1}\alpha\beta$?

7 If τ is a translation of the plane and β is any isometry, how many fixed points does $\beta^{-1}\tau\beta$ have?

By expressing τ as a product of two reflections, show that $\beta^{-1}\tau\beta$ may also be expressed as a product of two reflections, and deduce that $\beta^{-1}\tau\beta$ is a translation.

8 If R is the group of rotations of the plane with centre 0 and τ is the translation mapping 0 to 1, show that every member of the set $\tau^{-1}R\tau$ fixes 1 and is in fact a rotation. Illustrate.

9 Two elements x, y of a group G are said to be *conjugate* when there exists a $g \in G$ such that $x = g^{-1}yg$.

In the Euclidean group of the plane, what can you say about the isometries which are conjugate to
(i) a reflection,
(ii) a rotation,
(iii) a translation,
(iv) a glide-reflection?

10 Exhibit the mapping of the elements of the group $D_4 = \langle a, b \rangle$ of qn 6.40 given by

$$g \mapsto a^{-1}ga, \text{ for the given } a \in D_4.$$

If D_4 is thought of as the group of symmetries of a square, name the isometries and their images under this mapping.

Exhibit also the mapping of D_4 given by $g \mapsto b^{-1}gb$, and again name the corresponding isometries and their images.

Conjugacy classes

11 May a nonidentity element ever be conjugate to the identity element in a group?

12 In any group, prove that $(g^{-1}ag)^n = g^{-1}a^ng$. Must two conjugate elements of a group have the same order?

13 Prove that conjugacy is an equivalence relation on a group.
The equivalence classes are called *conjugacy classes*.

14 Find the conjugacy classes in D_4. Use qn 12 to narrow down the possibilities. Use the images of a^2 in qn 10 to show that this element is in a singleton class. Find $g^{-1}bg$ for all $g \in D_4$ to determine the conjugacy class of b. Use qn 10 to complete the search.

15 If a group element belongs to a singleton conjugacy class, what may be said about its relationship with the other group elements?

16 If G is an abelian group, describe its conjugacy classes.

17 Must the elements in the centre of a group belong to singleton conjugacy classes? Does this characterise them?

18 Use qn 12 to prove that if a group contains a unique element of order 2, then that element lies in the centre of the group.

19 If $\{a, b, c, d\} = \{1, 2, 3, 4\}$ and $\alpha = \begin{pmatrix} 1 & 2 & 3 & 4 \\ a & b & c & d \end{pmatrix}$ is an element of S_4, find the images of a, b, c and d under the permutation $\alpha^{-1}(1234)\alpha$ and write this permutation in cycle form. Is every element of S_4 conjugate to (1234) necessarily a 4-cycle? Is every 4-cycle in S_4 necessarily conjugate to (1234)?

20 By examining the permutation $\alpha^{-1}(123)\alpha$ determine whether every element of S_4 which is conjugate to (123) is necessarily a 3-cycle and whether every 3-cycle in S_4 is necessarily conjugate to (123).

21 Use qn 2 to determine the conjugacy class of (12) in S_4.

22 Find the conjugacy class of (12)(34) in S_4.

23 Let $\alpha = (1234)$ and $\beta = (12)$, find $\gamma = \alpha^{-1}\beta\alpha$, $\delta = \gamma^{-1}\beta\gamma$ and $\alpha^{-2}\gamma\alpha^2$. Use qn 6.46 to prove that α and β generate S_4.

Normal subgroups and conjugacy classes

24 If K is the kernel of the group homomorphism $\alpha: G \to H$ and $k \in K$, prove that $g^{-1}kg \in K$ for all $g \in G$. Deduce that if N is a normal subgroup of a group G, then $g^{-1}Ng \subseteq N$ for all $g \in G$.

25 If N is a subgroup of a group G and $g^{-1}Ng \subseteq N$ for *all* $g \in G$, prove that $Ng \subseteq gN$ and $g^{-1}N \subseteq Ng^{-1}$, so that $Ng = gN$ and N is a normal subgroup of G.

Questions 24 and 25 together establish the very important result that a subgroup N of a group G is a normal subgroup if and only if $g^{-1}Ng \subseteq N$ for all $g \in G$.

26 Prove that a normal subgroup of a group is a union of conjugacy classes.

27 By seeing what each conjugacy class of D_4 generates (qn 14) determine the normal subgroups of D_4.
What groups may be homomorphic images of D_4?

28 (i) If N is a normal subgroup of S_4 and N contains a transposition, use qns 21 and 6.45 to prove that $N = S_4$.
(ii) Express the product $(abdc)\,(adbc)\,(adcb)$ as a single cycle, and

deduce that if a normal subgroup N of S_4 contains a 4-cycle, then $N = S_4$.

(iii) If N is a normal subgroup of S_4 and N contains a 3-cycle, use qn 6.47 to prove that $N \supseteq A_4$. Since A_4 is a subgroup of S_4 of index 2, A_4 is a normal subgroup.

(iv) Find a normal subgroup of order 4 in S_4.

(v) What groups may be homomorphic images of S_4?

Direct products

29 If H and K are normal subgroups of a group G and $h \in H$ and $k \in K$, prove that $h^{-1}k^{-1}hk \in H \cap K$. If, further, $H \cap K = \{e\}$, prove that $hk = kh$.

30 Verify that the conditions of qn 10.11 require both the subgroups A and B to be normal subgroups of G.

31 Let $G = \{1, 3, 5, 7, 9, 11, 13, 15\}$, then G is a group in $(\mathbb{Z}_{16}, \cdot)$. Find the order of each of the elements in this group. Does $G = HK$ when

(i) $H = \{1, 7\}$ and $K = \{1, 15\}$,

(ii) $H = \{1, 3, 9, 11\}$ and $K = \{1, 15\}$,

(iii) $H = \{1, 3, 9, 11\}$ and $K = \{1, 5, 9, 13\}$?

32 If H and K are normal subgroups of G, $G = HK$ and $H \cap K = \{e\}$, use qns 10.11 and 29 to prove that G is isomorphic to the direct product $H \times K$.

33 Show that the group G of qn 31 is isomorphic to a direct product of two cyclic groups.

Centralisers

34 Use the table given to find the subsets of elements of D_4 which commute with each element of the group.

	e	a	a^2	a^3	b	ba	ba^2	ba^3
e	e	a	a^2	a^3	b	ba	ba^2	ba^3
a	a	a^2	a^3	e	ba^3	b	ba	ba^2
a^2	a^2	a^3	e	a	ba^2	ba^3	b	ba
a^3	a^3	e	a	a^2	ba	ba^2	ba^3	b
b	b	ba	ba^2	ba^3	e	a	a^2	a^3
ba	ba	ba^2	ba^3	b	a^3	e	a	a^2
ba^2	ba^2	ba^3	b	ba	a^2	a^3	e	a
ba^3	ba^3	b	ba	ba^2	a	a^2	a^3	e

The subset which commutes with a $\{c | ca = ac\}$ is called the *centraliser of* a and is denoted by C_a.

35 If g is a given element of a group G, prove that the centraliser of g, C_g, is necessarily a subgroup of G.

36 Compare the orders of the centralisers of the elements of D_4 with the size of the conjugacy classes containing these elements (qn 14).

37 If in a group G, $x^{-1}ax = b = y^{-1}ay$, prove that x and y belong to the same right coset of C_a. Prove also that if x and y belong to the same right coset of C_a, and $x^{-1}ax = b$, then $y^{-1}ay = b$. Deduce that there is a well-defined bijection

$$C_a x \mapsto x^{-1}ax$$

of the right cosets of C_a onto the conjugacy class containing a. If G is a finite group, how does the size of the conjugacy class containing a relate to the subgroup C_a? Prove that the size of a conjugacy class divides the order of G.

Normal subgroups of A_4

38 Let α be a 3-cycle in S_4. How many elements are there in the conjugacy class containing α? Use qn 37 to determine the order of C_α in S_4, and identify its elements. What is the centraliser of α in A_4? Use qn 37 to determine the size of the conjugacy class of α in A_4.
Express the products (132)(12)(34)(123), (123)(12)(34)(132), (12)(34)(123)(12)(34), (13)(24)(123)(13)(24) and (14)(23)(123)(14)(23) as disjoint cycles and list the conjugacy classes of A_4. Notice that they must be subsets of the conjugacy classes of S_4.

39 If N is a normal subgroup of A_4 and N contains a 3-cycle, prove that N contains all the 3-cycles and so $N = A_4$. Deduce that the only nontrivial normal subgroup of A_4 has order 4.

40 Why is it impossible for A_4 to have a subgroup of order 6? (This makes the converse of Lagrange's theorem false.)

Normal subgroups of A_5

41 If $\{1, 2, 3, 4, 5\} = \{a, b, c, d, e\}$, prove that either $\begin{pmatrix} 1 & 2 & 3 & 4 & 5 \\ a & b & c & d & e \end{pmatrix}$ or $\begin{pmatrix} 1 & 2 & 3 & 4 & 5 \\ a & b & c & e & d \end{pmatrix}$ is an even permutation. Deduce that A_5 is triply transitive and prove that the 3-cycles form a single conjugacy class in A_5.

42 Let α be a 3-cycle in S_5. What is the number of elements in the conjugacy class of α? Use qn 37 to determine the order of C_α in S_5, and identify the elements of $C_{(123)}$. What is the centraliser of (123) in A_5? Use qn 37 to determine the size of the conjugacy class of (123) in A_5, and confirm the result of qn 41.

43 List the cycle types in S_5 and say which give even and which give odd permutations.

44 If N is a normal subgroup of A_5 and N contains a 3-cycle use qns 41 and 6.47 to prove that $N = A_5$.

45 If N is a normal subgroup of A_5 and N contains an element $(ab)(cd)$ of order 2, explain why the product $(ab)(cd)[\alpha^{-1}(ab)(cd)\alpha]$, where $\alpha = (ab)(de)$, also gives an element of N. Deduce that N contains a 3-cycle, so that, by qn 44, $N = A_5$.

46 If N is a normal subgroup of A_5 and N contains an element $(abcde)$ of order 5, explain why the product $(abcde)[\alpha^{-1}(abcde)\alpha]$, where $\alpha = (bde)$, also gives an element of N. Deduce that N contains a 3-cycle, so that, by qn 44, $N = A_5$.

47 Use qns 44, 45 and 46 to show that the only normal subgroups of A_5 are A_5 itself and the subgroup consisting of the identity.

When the only normal subgroups of a group G are G itself and the singleton identity, the group G is said to be a *simple* group. Which cyclic groups are simple?

48 If $\alpha = (12345)$ and $\beta = (123)$, calculate $\beta(\alpha^{-1}\beta\alpha)$ and deduce three distinct prime factors of the order of the group $\langle \alpha, \beta \rangle$. Explain why A_5 may not have a subgroup of order 30. Prove that $\langle \alpha, \beta \rangle = A_5$.

49 Let α be a 5-cycle in S_5. How many elements are there in the conjugacy class containing α? Use qn 37 to determine the order of C_α in S_5, and identify its elements. What is the centraliser of α in A_5? Use qn 37 to determine the size of the conjugacy class of α in A_5.

Conjugate rotations in three dimensions

50 Let α and β be rotations of $\mathbb{R}^3$ and let α be a rotation through an angle θ about an axis a. If l is a line perpendicular to a describe the position of $l\alpha$.
Describe also the relation between the three lines $a\beta$, $l\beta$ and $l\alpha\beta$. Verify that $\beta^{-1}\alpha\beta$ fixes every point on $a\beta$ and maps $l\beta$ to $l\alpha\beta$. Deduce that $\beta^{-1}\alpha\beta$ is a rotation through an angle θ about the axis $a\beta$.

51 Use qns 50 and 5.14 to distinguish geometrically between the two conjugacy classes of 5-cycles in A_5.

Automorphisms

52 By computing the product $(a^{-1}g_1a)\,(a^{-1}g_2a)$ prove that the function $g \mapsto a^{-1}ga$ of qn 10 is structure-preserving and is an isomorphism of the group D_4 onto itself.

53 If g is any element of a group G, prove that the function $G \to G$ given by $x \mapsto g^{-1}xg$ is an isomorphism of the group onto itself.

Any isomorphism of a group onto itself is known as an *automorphism* of the group.

54 Under an automorphism of a group, must every subgroup be mapped onto a subgroup?

55 Under an automorphism of a group, must every conjugacy class be mapped onto a conjugacy class?

56 Under an automorphism of a group, must every normal subgroup be mapped onto a normal subgroup?

57 An automorphism constructed in the manner given in qn 53 is known as an *inner automorphism*. Under an inner automorphism, must each conjugacy class be mapped onto itself?

58 Under an inner automorphism, must each normal subgroup be mapped onto itself?

59 Prove that the function given by

$\alpha \mapsto (12)\alpha(12)$

is an automorphism of the group A_4.
Check that this automorphism is not given in the form of an inner automorphism. By examining the image of a conjugacy class of 3-cycles in A_4, prove that this automorphism is not an inner automorphism.

An automorphism of a group which is *not* an inner automorphism is known as an *outer automorphism*.

60 Prove that the function given by

$g \mapsto g^{-1}$

gives an automorphism of any abelian group, and that unless the function is the identity (as it is for the group C_2) it is an outer automorphism.

61 Do the inner automorphisms of a group G form a subgroup of S_G?

62 Do the automorphisms of a group G form a subgroup of S_G?

Summary

Definition Two elements x and y of a group G are said to be *con-*
qn 9   *jugate* when there exists a $g \in G$ such that $x = g^{-1}yg$.

Theorem Conjugacy is an equivalence relation on a group.
qn 13

Theorem Each conjugacy class of S_n contains all the permutations
qns 19–22 with a given cycle structure.

Theorem If N is a subgroup of a group G, then N is a normal
qns 24, 25 subgroup of G if and only if $g^{-1}Ng \subseteq N$ for all $g \in G$.

Theorem A normal subgroup is a union of conjugacy classes.
qn 27

Theorem If H and K are normal subgroups of a group G and
qn 32 $G = HK$ and $H \cap K = \{e\}$, then G is isomorphic to
        the direct product $H \times K$.

Definition The subset of elements of a group G which commute
qn 34   with a given element a of G is called the *centraliser* of a
        in G and denoted by C_a.

Theorem The centraliser of a in G is a subgroup of G and if G is
qn 37   a finite group the index of C_a in G is equal to the num-
        ber of elements in the conjugacy class of a.

Theorem The only normal subgroups of A_5 have orders 1 and 60.
qn 47

Theorem If a is a given element of a group G, the function given
qn 57   by $g \mapsto a^{-1}ga$ is an isomorphism of the group onto
        itself called an *inner automorphism*.

Theorem The inner automorphisms of a group G form a sub-
qn 61   group of S_G.

Theorem The set of all isomorphisms of a group G onto itself is a
qn 62   subgroup of S_G called the automorphism group of G.

Historical note

In 1845 A. L. Cauchy established that the conjugacy classes of S_n
were the sets of permutations with the same cycle structure. F. Klein's
Lectures on the Icosahedron (1884) make use of conjugacy of geometrical transformations, especially rotations, of three dimensional space.

Inner and outer automorphisms were distinguished by F. G.
Frobenius in 1901.

Answers to chapter 17

1 (23), (23).

2 $\alpha^{-1}(12)\alpha = (1\alpha 2\alpha)$.

3 Stabiliser of 4 mapped to stabiliser of 1.

4 $(m\beta)\beta^{-1}\alpha\beta = m\beta \Leftrightarrow m\alpha\beta = m\beta \Leftrightarrow m\alpha = m$.

5 The unique fixed point makes it a rotation. See qn 3.31.

6 If l is the axis of α, $\beta^{-1}\alpha\beta$ fixes every point on $l\beta$ and so is a reflection. See qn 3.31.

7 No fixed points. If $\tau = \varrho_1\varrho_2$, then $\beta^{-1}\tau\beta = \beta^{-1}\varrho_1\varrho_2\beta = (\beta^{-1}\varrho_1\beta)(\beta^{-1}\varrho_2\beta)$.

8 $1\tau^{-1}R\tau = 0R\tau = 0\tau = 1$. Fixed point unique.

9 (i) Only reflections from qn 6. (ii) Only rotations from qn 5. (iii) Only translations from qn 7. (iv) Only glide-reflections: no fixed points and if
$\gamma = \varrho_1\varrho_2\varrho_3$ $\beta^{-1}\gamma\beta = (\beta^{-1}\varrho_1\beta)(\beta^{-1}\varrho_2\beta)(\beta^{-1}\varrho_3\beta)$.   See qn 6.49.

10 $\begin{pmatrix} g \\ a^{-1}ga \end{pmatrix} = \begin{pmatrix} e & a & a^2 & a^3 & b & ba & ba^2 & ba^3 \\ e & a & a^2 & a^3 & ba^2 & ba^3 & b & ba \end{pmatrix}$

$\begin{pmatrix} g \\ b^{-1}gb \end{pmatrix} = \begin{pmatrix} e & a & a^2 & a^3 & b & ba & ba^2 & ba^3 \\ e & a^3 & a^2 & a & b & ba^3 & ba^2 & ba \end{pmatrix}$

11 Never.

12 $(g^{-1}ag)(g^{-1}ag) \ldots (g^{-1}ag) = g^{-1}aa \ldots a\,g$.
       n times n times

13 Reflexive: $a = e^{-1}ae$. Symmetric: $b = g^{-1}ag \Rightarrow gbg^{-1} = a \Rightarrow a = (g^{-1})^{-1}bg^{-1}$. Transitive: $b = g^{-1}ag$ and $c = h^{-1}bh \Rightarrow c = h^{-1}(g^{-1}ag)h = (gh)^{-1}a(gh)$.

14 From qn 11, $\{e\}$ is one class. Only elements of order 4 are a, a^3 so $\{a, a^3\}$ is one class from qns 12 and 10. $b^{-1}a^2b = a^2$ and $a^{-1}a^2a = a^2$ so $(a^ib^j)^{-1}a^2(a^ib^j) = a^2$. So $\{a^2\}$ is a singleton class. $g^{-1}bg = b$ for $g = e, a^2$, b, ba^2. $g^{-1}bg = ba^2$ for $g = a, a^3, ba, ba^3$. So $\{b, ba^2\}$ is a class. The remaining two elements are known to be conjugate from qn 10.

15 If $g^{-1}ag = a$ for all g, then $ag = ga$ and a is in the centre.

16 All singletons.

17 Yes.

18 If a is the unique element of order 2, $g^{-1}ag = a$ from qn 12.

19 $(abcd)$. Yes. Yes.

20 With α as in qn 19, $\alpha^{-1}(123)\alpha = (abc)$.

21 All transpositions in one conjugacy class.

22 $\{(12)(34), (13)(24), (14)(23)\}$.

23 $\gamma = (23)$, $\delta = (13)$, $\alpha^{-2}\gamma\alpha^2 = (14)$.

24 $(g^{-1}kg)\alpha = g^{-1}\alpha \cdot e\alpha \cdot g\alpha = g^{-1}\alpha \cdot g\alpha = (g^{-1}g)\alpha = e\alpha$. Every normal subgroup is the kernel of a homomorphism, from qn 16.22.

25 $g^{-1}Ng \subseteq N \Rightarrow g(g^{-1}Ng) \subseteq gN$. Also $g^{-1}Ng \subseteq N \Rightarrow (g^{-1}Ng)g^{-1} \subseteq Ng^{-1}$.

26 Since $g^{-1}Ng \subseteq N$, every conjugate of an element of N is in N.

27 $\langle e \rangle = \{e\}$, $\langle a^2 \rangle = \{e, a^2\}$, $\langle a, a^3 \rangle = \{e, a, a^2, a^3\}$, $\langle b, ba^2 \rangle = \{e, b, ba^2, a^2\}$, $\langle ba, ba^3 \rangle = \{e, ba, ba^3, a^2\}$. Each of these is a union of conjugacy classes. Any other pair of classes generate the whole group. If H has order 4, D_4/H has order 2 and is isomorphic to C_2. $D_4/\langle a^2 \rangle$ was exhibited in qn 16.21, isomorphic to D_2.

28 (i) From qn 21, N contains all transpositions. (ii) (ab). If N contains one 4-cycle, then by qn 19 all 4-cycles and hence a transposition. (iii) From qn 20, N contains all 3-cycles and hence A_4. (iv) (1), (12)(34), (13)(24), (14)(23) contains two complete conjugacy classes. This subgroup is sometimes known as V_4. (v) $S_4/A_4 \cong C_2$. $S_4/V_4 \cong S_3 \cong D_3$.

29 H normal $\Rightarrow k^{-1}hk \in H \Rightarrow h^{-1}k^{-1}hk \in H$. K normal $\Rightarrow h^{-1}k^{-1}h \in K \Rightarrow h^{-1}k^{-1}hk \in K$.

30 From (i) $g^{-1}ag = (a_1b_1)^{-1}a(a_1b_1) = b_1^{-1}a_1^{-1}aa_1b_1 = a_1^{-1}aa_1$ from (ii).

31 Order 1: 1, order 2: 7, 9, 15, order 4: 3, 5, 11, 13. (i) No. (ii) Yes. (iii) Yes.

32 $h_1k_1 = h_2k_2 \Rightarrow h_2^{-1}h_1 = k_2k_1^{-1}$, so qn 10.11 (i) follows from $H \cap K = \{e\}$. H, K normal with $H \cap K = \{e\}$ gives qn 10.11 (ii) from qn 29.

33 Qn 31 (ii) gives $HK = G$, $H \cap K = \{e\}$. $H \cong C_4$. $K \cong C_2$.

34 $C_a = \{e, a, a^2, a^3\}$. $C_b = \{e, b, a^2, ba^2\}$. $C_{a^2} = D_4$.

35 Check closure, identity, inverses.

36

Conjugacy class	Centraliser
$\{e\}$	D_4
$\{a^2\}$	D_4
$\{a, a^3\}$	$\{e, a, a^2, a^3\}$.
$\{b, ba^2\}$	$\{e, b, ba^2, a^2\}$.
$\{ba, ba^3\}$	$\{e, ba, ba^3, a^2\}$.

(Size of class) $\times$ (order of centraliser) $= 8$.

37 $x^{-1}ax = y^{-1}ay \Leftrightarrow yx^{-1}a = ayx^{-1} \Leftrightarrow yx^{-1} \in C_a$. Thus $C_ax = C_ay \Leftrightarrow x^{-1}ax = y^{-1}ay$. So number of cosets of $C_a =$ number of elements conjugate to a. Index of C_a divides $|G|$.

38 A 3-cycle α has eight conjugates in S_4 so C_α has order 3. Thus $C_\alpha = \{I, \alpha, \alpha^2\}$. These are all in A_4 so α has four conjugates in A_4. $\{I\}$, $\{(12)(34), (13)(24), (14)(23)\}$, $\{(123), (214), (341), (432)\}$, $\{(132), (241), (314), (423)\}$.

39 The conjugacy class of one 3-cycle contains four 3-cycles. The inverses of these four belong to another conjugacy class.

40 If A_4 had a subgroup of order 6, it would have index 2 and thus be normal, contradicting qn 39.

41 $\begin{pmatrix} 1 & 2 & 3 & 4 & 5 \\ a & b & c & d & e \end{pmatrix} (de) = \begin{pmatrix} 1 & 2 & 3 & 4 & 5 \\ a & b & c & e & d \end{pmatrix}$, so one is even and one is odd. So there is an even permutation in S_5 such that $1 \mapsto a$, $2 \mapsto b$ and $3 \mapsto c$. This makes A_5 triply transitive. If α is the even permutation of these $\alpha^{-1}(123)\alpha = (abc)$ so all 3-cycles are conjugates in A_5.

42. 20. The order of C_α in S_5 is 6.
$C_{(123)} = \{(1), (123), (132), (45), (123)(45), (132)(45)\}$ in S_5. $C_{(123)} \cap A_5 = \{(1), (123), (132)\}$. The conjugacy class of (123) in A_5 contains 60/3 elements, i.e. all 3-cycles.

43 Even: $(a)(b)(c)(d)(e)$; $(a)(bc)(de)$; $(a)(b)(cde)$; $(abcde)$.
Odd: $(a)(b)(c)(de)$; $(a)(bcde)$; $(ab)(cde)$.

45 Let $\beta = (ab)(cd)$, then $\alpha^{-1}\beta\alpha \in N$ since N is normal and $\beta\alpha^{-1}\beta\alpha \in N$ since N is closed. $\beta\alpha^{-1}\beta\alpha = (cde)$.

46 Argue as in qn 45. (bed).

47 There are elements of three cycle types in A_5 distinct from the identity. The last three questions have dealt with the three types, and show that any one in a normal subgroup N of A_5 implies $N = A_5$. Cyclic groups of prime order are simple.

48 $\beta\alpha^{-1}\beta\alpha = (13)(24)$. $\langle\alpha, \beta\rangle$ contains elements of order 2, 3 and 5 so order of $\langle\alpha, \beta\rangle = 2\cdot3\cdot5\cdot k = 30k$. But a subgroup of index 2 would be normal, so $k = 2$.

49 There are 24 5-cycles in S_5, so centraliser of α has index 24 and so order 5. Thus $C_\alpha = \{I, \alpha, \alpha^2, \alpha^3, \alpha^4\}$. These are all even permutations. So in A_5, C_α has order 5 and so index 12. Conjugacy class of α in A_5 contains 12 elements.

50 The line $l\alpha$ is also perpendicular to a and is inclined at an angle θ to l. The lines $l\beta$ and $l\alpha\beta$ are both perpendicular to $a\beta$ and inclined to each other at an angle θ. Now $A \in a \Rightarrow A\beta \in a\beta$ and $A\beta(\beta^{-1}\alpha\beta) = A\beta$ so $a\beta$ is the line of fixed points of $\beta^{-1}\alpha\beta$. Moreover, $l\beta(\beta^{-1}\alpha\beta) = l\alpha\beta$, so $\beta^{-1}\alpha\beta$ is a rotation with axis $a\beta$ through an angle θ.

51 The conjugate of a rotation is a rotation through the same angle. Thus rotations through $\pm\ 72°$ may be conjugate to each other, but not conjugate to rotations through $\pm\ 144°$.

53 $g^{-1}xg = g^{-1}yg \Rightarrow x = y$ so mapping is one–one. $g^{-1}xyg = (g^{-1}xg)\,(g^{-1}yg)$ so mapping is structure-preserving.

54 Yes, by qn 16.11.

55 $b = x^{-1}ax \Rightarrow b\alpha = (x^{-1})\alpha \cdot a\alpha \cdot x\alpha = (x\alpha)^{-1}a\alpha\,(x\alpha)$ for any automorphism α.

56 A normal subgroup is a union of conjugacy classes. Yes.

57 Yes.

58 Yes.

59 If α is an even permutation, $(12)\alpha(12)$ is even. $(12)\alpha(12) = (12)\beta(12) \Rightarrow$ $\alpha = \beta$. $(12)\alpha\beta(12) = [(12)\alpha(12)]\,[(12)\beta(12)]$. If $\alpha = (123)$, $(12)\alpha(12) = (213)$ which is not in the same conjugacy class as (123) in A_4, so by qn 57 this is not an inner automorphism.

60 For an abelian group, every inner automorphism is the identity map.

61 Composite of $x \mapsto g_1^{-1}xg_1$ and $x \mapsto g_2^{-1}xg_2$ is

$$x \mapsto g_2^{-1}(g_1^{-1}xg_1)g_2 = (g_1g_2)^{-1}x(g_1g_2)$$

so the set of inner automorphisms is closed. $g_1 = e$ gives identity map. $g_2 = g_1^{-1}$ gives inverse map.

62 Automorphisms of G are elements of S_G by definition. If $(xy)\alpha = x\alpha \cdot y\alpha$ and $(xy)\beta = x\beta \cdot y\beta$ then $(xy)\alpha\beta = [x\alpha \cdot y\alpha]\beta = x\alpha\beta \cdot y\alpha\beta$ so the set of automorphisms is closed. The identity map is an automorphism. $(x\alpha^{-1} \cdot y\alpha^{-1})\alpha = xy$ so $x\alpha^{-1} \cdot y\alpha^{-1} = (xy)\alpha^{-1}$.

18

Linear fractional groups

If we examine the effect of the elements of the general linear group on 1-dimensional subspaces we obtain a transitive permutation group which is a homomorphic image of the general linear group. This provides a setting in which the Möbius group and the projective group on a line may be seen as examples of an infinite class of triply transitive groups. In the development of this chapter, the 1-dimensional subspaces of $V_2(F)$ become the set of objects being permuted in a permutation group. This is the step by which we move from a vector space to a projective space, in this case from a 2-dimensional vector space to a projective line. The 1-dimensional subspaces are the points of the projective line. More generally the 1-dimensional subspaces of $V_n(F)$ form an $(n-1)$-dimensional projective space. Because different vectors, e.g. (x, y) and (kx, ky), lie in the same 1-dimensional subspace, either pair of coordinates uniquely determines a projective point provided $k \neq 0$. In such circumstances we sometimes talk of using homogeneous coordinates for the projective points.

Permutations of 1-dimensional subspaces

1 Let

$$\alpha : (x, y) \mapsto (x, y) \begin{pmatrix} 2 & 3 \\ 4 & 5 \end{pmatrix}$$ be a linear transformation of $V_2(\mathbb{R})$. Find

the images of the points $(2, 1)$, $(4, 2)$, $(6, 3)$ and generally $(2k, k)$ under this transformation. Do the images of these points all lie in the same 1-dimensional subspace? If the subspace $\{x(2, 1) | x \in \mathbb{R}\}$ is mapped to the subspace $\{x(s, 1) | x \in \mathbb{R}\}$ under α, what is the number s?

Is the subspace $\{x(3, 1) | x \in \mathbb{R}\}$ mapped to a subspace $\{x(t, 1) | x \in \mathbb{R}\}$ under α? If so, find t.

For a given real number m, must the subspace $\{x(m, 1) | x \in \mathbb{R}\}$ be mapped to a subspace of $V_2(\mathbb{R})$ under α? Supposing that $3m + 5 \neq 0$, can you find a number m' in terms of m such that $\{x(m, 1)|x \in \mathbb{R}\}$ is mapped to $\{x(m', 1)|x \in \mathbb{R}\}$ under α?

2 Let

$$\beta : (x, y) \mapsto (x, y)\begin{pmatrix} 2k & 3k \\ 4k & 5k \end{pmatrix}$$

be a linear transformation of $V_2(\mathbb{R})$, and let $k \neq 0$. Find the image of the subspace $\{x(m, 1)|x \in \mathbb{R}\}$ under the transformation β, and again, presuming that $3m + 5 \neq 0$, find a number m' in terms of m such that $\{x(m, 1)|x \in \mathbb{R}\}$ is mapped to $\{x(m', 1)|x \in \mathbb{R}\}$ under β.

3 Is every 1-dimensional subspace of $V_2(\mathbb{R})$ either of the form $\{x(m, 1)|x \in \mathbb{R}\}$ for some suitable choice of m, or of the form $\{(x, 0)|x \in \mathbb{R}\}$?

If the 1-dimensional subspaces of $V_2(\mathbb{R})$ are labelled with the elements of the set $\mathbb{R} \cup \{\infty\}$ as follows

$\{x(m, 1)|x \in \mathbb{R}\} \mapsto m,$
$\{(x, 0)|x \in \mathbb{R}\} \mapsto \infty,$

exhibit the permutation of $\mathbb{R} \cup \{\infty\}$ which is induced by α (the α of qn 1). State explicitly which m is mapped to ∞, and what the image of ∞ is.

4 Is every 1-dimensional subspace of $V_2(\mathbb{Z}_3)$ either of the form $\{x(m, 1)|x \in \mathbb{Z}_3\}$ for some suitable choice of m, or of the form $\{(x, 0)|x \in \mathbb{Z}_3\}$?

Find the image of each of the subspaces $\{x(0, 1)|x \in \mathbb{Z}_3\}$, $\{x(1, 1)|x \in \mathbb{Z}_3\}$, $\{x(2, 1)|x \in \mathbb{Z}_3\}$ and $\{(x, 0)|x \in \mathbb{Z}_3\}$ under the linear transformation

$$\alpha : (x, y) \mapsto (x, y)\begin{pmatrix} 0 & 1 \\ 1 & 1 \end{pmatrix}.$$

Does every nonsingular linear transformation of $V_2(\mathbb{Z}_3)$ necessarily permute the 1-dimensional subspaces?

If the 1-dimensional subspaces of $V_2(\mathbb{Z}_3)$ are labelled with the elements of the set $\mathbb{Z}_3 \cup \{\infty\}$ as follows

$\{x(m, 1)|x \in \mathbb{Z}_3\} \mapsto m,$
$\{(x, 0)|x \in \mathbb{Z}_3\} \mapsto \infty,$

exhibit the permutation of $\mathbb{Z}_3 \cup \{\infty\}$ which is induced by α. Give a general algebraic expression for the image of m, provided $m + 1 \neq 0$? Write down another linear transformation of $V_2(\mathbb{Z}_3)$ which permutes the 1-dimensional subspaces in exactly the same way as α.

5 If α and β are nonsingular linear transformations of $V_2(F)$ which permute the 1-dimensional subspaces in exactly the same way, what is the effect of $\alpha\beta^{-1}$ on the 1-dimensional subspaces of $V_2(F)$?

If $(1, 0)\alpha\beta^{-1} = (s, 0)$ and $(0, 1)\alpha\beta^{-1} = (0, t)$, prove that
$(1, 1)\alpha\beta^{-1} = (s, t)$ and deduce that $s = t$. If $\alpha\beta^{-1}:(x, y) \mapsto (x, y)A$,
what kind of matrix is A?

6 Treating $GL(2, \mathbb{Z}_3)$ as a group of matrices, find the centre of this group
using qn 13.16. Write down three distinct cosets of this centre, and
find the permutations induced on the 1-dimensional subspaces of
$V_2(\mathbb{Z}_3)$ by the six transformations derived from these matrices.

7 If under the nonsingular linear transformation of $V_2(F)$

$$\alpha:(x, y) \mapsto (x, y) \begin{pmatrix} a & c \\ b & d \end{pmatrix}$$

the 1-dimensional subspace $\{x(m, 1)|x \in F\}$ is mapped to the
1-dimensional subspace $\{x(m', 1)|x \in F\}$, assuming that
$cm + d \neq 0$, express m' in terms of m. What is the image of the
subspace $\{x(-d, c)|x \in F\}$ under α? What is the image of the sub-
space $\{(x, 0)|x \in F\}$ under α?

8 If we attempt to define a mapping $F \to F$ by

$$x \mapsto \frac{ax + b}{cx + d},$$

where $a, b, c, d \in F$ and $c \neq 0$, what element of F must be deleted
from the domain?

9 Use the equation

$$\frac{ax + b}{cx + d} = \frac{b}{d} + \frac{ad - bc}{d(cx + d)} x$$

to show that when $ad - bc = 0$, the mapping

$$x \mapsto \frac{ax + b}{cx + d}$$

is not an injection of the domain $F - \{-d/c\}$.

10 If $ad - bc \neq 0$ and $c \neq 0$, find the field element s if the mapping given
by

$$x \mapsto \frac{ax + b}{cx + d}$$

is a bijection of $F - \{-d/c\}$ to $F - \{s\}$.

11 If for each $m \in F$, a nonsingular linear transformation α of $V_2(F)$ maps
the 1-dimensional subspace $\{x(m, 1)|x \in F\}$ onto the 1-dimensional
subspace $\{x(am + b, 1)|x \in F\}$ where $a \neq 0$, what is the image
and the pre-image of the subspace $\{(x, 0)|x \in F\}$ under α?

12 When $ad - bc \neq 0$ we define a *linear fractional transformation* of
$F \cup \{\infty\}$ by

$$\alpha:x \mapsto \frac{ax + b}{cx + d}$$

when $c = 0$, $\infty\alpha = \infty$, and when $c \neq 0$, $(-d/c)\alpha = \infty$ and

$\infty \alpha = a/c$. The set $F \cup \{\infty\}$ is called the *projective line* over the field F.

Write down the six linear fractional transformations of $\mathbb{Z}_2 \cup \{\infty\}$ and determine the corresponding permutations of $\{0, 1, \infty\}$.

The homomorphism $GL(2, F) \to LF(F)$

13 Prove that the mapping of the group $GL(2, F)$, of matrices under matrix multiplication, onto the set of linear fractional transformations under composition, given by

$$\begin{pmatrix} a & c \\ b & d \end{pmatrix} \mapsto \left[x \mapsto \frac{ax + b}{cx + d} \right]$$

is a homomorphism. Deduce that the linear fractional transformations form a group under composition. This group is denoted by $LF(F)$.

14 If in qn 13 we take $F = \mathbb{Z}_3$, can you distinguish between the images of the matrices $\begin{pmatrix} 1 & 2 \\ 0 & 1 \end{pmatrix}$ and $\begin{pmatrix} 2 & 1 \\ 0 & 2 \end{pmatrix}$ under the homomorphism?

15 What is the kernel of the homomorphism of qn 13?

16 From qn 13.32 determine the order of $GL(2, \mathbb{Z}_3)$ and of its centre C. Deduce the order of the group $LF(\mathbb{Z}_3)$.

17 Exhibit the transformations $x \mapsto 1/(x + 1)$ and $x \mapsto 1/x$ in $LF(\mathbb{Z}_3)$ as permutations of the set $\{0, 1, 2, \infty\}$ and deduce from qn 17.23 that $LF(\mathbb{Z}_3)$ is isomorphic to S_4.

18 With the notation of qn 11.8, determine the number of vectors in $V_2(F_4)$ and the number of vectors in a 1-dimensional subspace. Deduce the order of $GL(2, F_4)$. How many matrices are there in the centre of this group?
Use these results to determine the order of $LF(F_4)$, by applying the fundamental theorem on homomorphisms.

19 With the notation of qn 11.8, exhibit the transformations $x \mapsto b/(x + b)$ and $x \mapsto bx + 1$ in $LF(F_4)$ as permutations of the set $\{0, 1, a, b, \infty\}$ and deduce from qn 17.48 that $LF(F_4)$ is isomorphic to A_5.

The quotient group $PGL(2, F)$

20 Apply the fundamental theorem on homomorphisms to qn 13 to establish that if C is the centre of the group $GL(2, F)$, then the quotient group

$$\frac{GL(2, F)}{C}$$

is isomorphic to the group $LF(F)$.

The quotient group $\dfrac{GL(2, F)}{C}$

is known as the *projective general linear group PGL(2, F)*. Since $PGL(2, F) \cong LF(F)$, we will use the two names interchangeably.

21 If a, b and c are distinct elements of F, what are their images under the linear fractional transformation

$$x \mapsto \frac{x - a}{x - b} \cdot \frac{c - b}{c - a}?$$

Deduce that the linear fractional group is triply transitive on the projective line $F \cup \{\infty\}$.

22 Prove that the only linear fractional transformation to fix ∞, 0 and 1 is the identity. Deduce from qn 21 that a linear fractional transformation is uniquely determined by three points and their images.

23 If F is a field containing q elements, show that $LF(F)$ contains $(q + 1)q(q - 1)$ elements. Deduce from qn 20 that $GL(2, F)$ contains $(q + 1)q(q - 1)^2$ elements, and verify this directly.

24 What can you say about a linear transformation of $V_2(F)$ with three eigenvectors, none of which is a scalar multiple of the others?

25 (Optional) By labelling the faces of a cube with the elements of $\mathbb{Z}_5 \cup \{\infty\}$ in such a way that the linear fractional transformations $x \mapsto 2x$ and $x \mapsto 3/x$ both correspond to rotational symmetries of the cube, show that there is a subgroup of $PGL(2, \mathbb{Z}_5)$ which is isomorphic to S_4.

Projective special linear group $PSL(2, F)$

26 If C denotes the centre of the group $GL(2, F)$, use the working of qn 13.16 to show that the centre of the group $SL(2, F)$ is $SL(2, F) \cap C = C_S$.

27 If F is a finite field containing q elements and q is even, prove that the multiplicative group of F contains no elements of even order, so that $x = 1$ is the only solution of the equation $x^2 = 1$. Deduce that for such a field, the centre of the group $SL(2, F)$ consists of the identity alone.

28 If F is a finite field containing q elements and q is odd, prove that the additive group of F contains no elements of even order, so that $x = 0$ is the only solution of the equation $x + x = 0$. Deduce that $+1$ and -1 are distinct, so that $x^2 = 1$ has exactly two solutions. Show that for such a field, the centre of the group $SL(2, F)$ contains just two distinct elements.

29 For $F = \mathbb{Z}_2$, $\mathbb{Z}_3$, F_4 and $\mathbb{Z}_5$ find the orders of $GL(2, F)$, $SL(2, F)$, the centres of each of these two groups, and the quotient groups $GL(2, F)/C$ and $SL(2, F)/C_S$.

30 Let S denote $SL(2, F)$, C_S the centre of S and let π denote the homomorphism of qn 13. Use qn 16.11 to show that $S\pi$ is a subgroup of $LF(F)$. What is the kernel of the homomorphism $\pi : S \to S\pi$? Use the fundamental theorem on homomorphisms (qn 16.23) to establish that the quotient group S/C_S is isomorphic to $S\pi$.

The quotient group S/C_S is known as the *projective special linear group PSL(2, F)*.

31 With the notation of qn 30, suppose that A is an element of $GL(2, F)$ such that $A\pi \in S\pi$. Prove that for some element $B \in SL(2, F)$, AB^{-1} lies in the centre of $GL(2, F)$, and deduce that the determinant of A is a square.

32 If $ad - bc = r^2 \neq 0$, prove that

$$x \mapsto \frac{ax + b}{cx + d}$$

lies in $S\pi$ (using the notation of qn 30) by constructing a matrix in $SL(2, F)$ with the same image under π.

33 From qns 30, 31 and 32 deduce that $PSL(2, F)$ is isomorphic to the subgroup of $LF(F)$ consisting of those elements with determinant a square.

34 By examining the elements $x \mapsto x + 1$ and $x \mapsto 2/x$ of $LF(\mathbb{Z}_3)$, prove that $PSL(2, \mathbb{Z}_3)$ is isomorphic to A_4.

35 (Optional). Label the twelve faces of a regular dodecahedron with the six elements of $\mathbb{Z}_5 \cup \{\infty\}$ in such a way that $x \mapsto x + 1$ and $x \mapsto 4/(x + 4)$, as elements of $LF(\mathbb{Z}_5)$ correspond to rotational symmetries of the dodecahedron. Verify, either from qn 29 or from qn 5.15 that the group generated by these two linear fractional transformations has order 60 or less. From qn 5.5(iv) deduce that these rotations of the dodecahedron correspond to rotational symmetries of a regular icosahedron and from qn 5.14 to elements of A_5. From qn 17.48, prove that $PSL(2, \mathbb{Z}_5)$ is isomorphic to A_5.

36 Use qn 20 to prove that $PGL(2, \mathbb{C})$ is isomorphic to the Möbius group, and use qn 33 to prove that $PSL(2, \mathbb{C})$ is also isomorphic to the Möbius group.

37 Use qns 7.46 and 7.47 to show that there is a homomorphism of the group of nonsingular complex matrices of the form $\begin{pmatrix} a & -\bar{b} \\ b & \bar{a} \end{pmatrix}$ onto the group of rotations of a sphere.

Summary

Theorem qns 7, 12 The elements of $GL(2, F)$ permute the 1-dimensional subspaces of $V_2(F)$. If the 1-dimensional subspaces are labelled

$$\{x(m, 1)|x \in F\} \mapsto m,$$
$$\{(x, 0)|x \in F\} \mapsto \infty,$$

then the permutation induced by $\begin{pmatrix} a & c \\ b & d \end{pmatrix}$ in $GL(2, F)$ is precisely the *linear fractional transformation*

$$x \mapsto \frac{ax + b}{cx + d},$$

where ∞ is fixed if $c = 0$ and $-d/c$ is mapped to ∞ and ∞ is mapped to a/c if $c \neq 0$.

Theorem qn 13 The mapping of $GL(2, F)$ given by

$$\begin{pmatrix} a & c \\ b & d \end{pmatrix} \mapsto \left[x \mapsto \frac{ax + b}{cx + d}\right]$$

is a homomorphism with kernel the centre of $GL(2, F)$ onto the linear fractional group $LF(F)$.

Theorem qn 20 If C denotes the centre of $GL(2, F)$ then the quotient group $GL(2, F)/C$, known as the *projective general linear group PGL(2, F)*, is isomorphic to the linear fractional group $LF(F)$.

Theorem qns 21, 22 The linear fractional group is triply transitive on the projective line $F \cup \{\infty\}$ and each transformation is uniquely determined by its action on three distinct elements.

Definition qn 30 If C_S denotes the centre of $SL(2, F)$ then the quotient group $SL(2, F)/C_S$ is known as the *projective special linear group PSL(2, F)*.

Theorem qn 33 $PSL(2, F)$ is isomorphic to the subgroup of linear fractional transformations for which $ad - bc$ is a square.

Further reading: Rotman, chapter 8; Carmichael, sections 68 and 71.

Historical note

The linear fractional groups for different fields arose independently. We have already seen, how, for the field of complex numbers, it was studied synthetically by A. F. Möbius (1852–56). For the field of real numbers, it appeared in the work of von Staudt (1847) as the projective group on a line, with elements formed by a sequence of projections from one line to another in the real projective plane.

For the field $\mathbb{Z}_p$, the linear fractional group and its subgroup with square determinants was studied by E. Galois (1832) who introduced the symbol ∞ with the meaning we attach to it in this chapter. For arbitrary finite fields, the linear fractional group was studied by E. H. Moore (1893) who established the simplicity of $PSL(2, F)$ for fields of order greater than 3. The homomorphism of $GL(2, F)$ to the linear fractional group is implied in the work of E. Galois (1832) and J. A. Serret (1866), and was used by A. Cayley (1880) to determine properties of linear fractional transformations.

Answers to chapter 18

1 $s = \frac{8}{11}$. $t = \frac{5}{7}$.

$$m' = \frac{2m + 4}{3m + 5}.$$

2 $m' = \frac{2m + 4}{3m + 5}.$

3 $m \mapsto \frac{2m + 4}{3m + 5}$, $-\frac{5}{3} \mapsto \infty$, $\infty \mapsto \frac{2}{3}$.

4 $\{x(0, 1)\} \to \{x(1, 1)\} \to \{x(2, 1)\} \to \{x(1, 0)\} \to \{x(0, 1)\}$. Yes from qns 12.5

and 13.5. $(0\ 1\ 2\ \infty)$. $m \mapsto 1/(m + 1)$. $\begin{pmatrix} 0 & 2 \\ 2 & 2 \end{pmatrix}$.

5 $\alpha\beta^{-1}$ fixes each 1-dimensional subspace. $(1, 1) = (1, 0) + (0, 1)$ so
$(1, 1)\alpha\beta^{-1} = (s, 0) + (0, t)$. If the subspace $\{x(1, 1) | x \in F\}$ is fixed by
$\alpha\beta^{-1}$, $s = t$. A is a scalar matrix.

6 Centre $= \left\{ \begin{pmatrix} 1 & 0 \\ 0 & 1 \end{pmatrix}, \begin{pmatrix} 2 & 0 \\ 0 & 2 \end{pmatrix} \right\}$. Cosets $\left\{ \begin{pmatrix} 0 & 1 \\ 1 & 1 \end{pmatrix}, \begin{pmatrix} 0 & 2 \\ 2 & 2 \end{pmatrix} \right\}, \left\{ \begin{pmatrix} 1 & 0 \\ 2 & 1 \end{pmatrix}, \begin{pmatrix} 2 & 0 \\ 1 & 2 \end{pmatrix} \right\}$

for example. Matrices in the same coset permute the 1-dimensional subspaces in the same way.

7 $m' = \frac{am + b}{cm + d}.$

$\{x(-d, c)\} \to \{(x, 0)\} \to \{x(a, c)\}.$

8 $-d/c.$

10 $s = a/c.$

11 $\{(x, 0)\}$ is fixed by qn 13.5.

12 (0): $x \mapsto x$,
(01∞): $x \mapsto 1/(x + 1)$,
$(0\infty 1)$: $x \mapsto (x + 1)/x$,
(01): $x \mapsto x + 1$,
(0∞): $x \mapsto 1/x$,
(1∞): $x \mapsto x/(x + 1)$.

13 The set of linear fractional transformations is a subset of $S_{F \cup \{\infty\}}$. Since this set
is the image of a group under a homomorphism it is a group by qn 16.11.

14 No.

15 The scalar matrices.

16 $|GL(2, \mathbb{Z}_3)| = 48$, $|\text{centre}| = 2$, so $|LF(\mathbb{Z}_3)| = 24$.

17 $(\infty 012): x \mapsto 1/(x + 1)$, $(\infty 0): x \mapsto 1/x$.

18 16 vectors in all. Four in a 1-dimensional subspace. $|GL(2, F_4)| = 15 \cdot 12 = 180$. Three matrices in centre. $|LF(F_4)| = 60$.

19 $(01ab\infty): x \mapsto b/(x + b)$, $(01a): x \mapsto bx + 1$.

20 C is the kernel of the homomorphism.

21 $a \mapsto 0$, $b \mapsto \infty$, $c \mapsto 1$. Argue as in qn 4.12.

22 To fix ∞, $c = 0$. To fix ∞ and 0, $b = c = 0$. To fix ∞, 0 and 1, $b = c = 0$ and $a = d$. If β and γ have the same effect on a, b and c and α is the transformation of qn 21 then $\alpha^{-1}(\beta\gamma^{-1})\alpha$ fixes 0, 1 and ∞. From qn 22, $\alpha^{-1}\beta\gamma^{-1}\alpha = 1$, so $\beta = \gamma$.

23 $LF(F)$ is triply transitive on $F \cup \{\infty\}$, i.e. $q + 1$ elements. So ∞ may map to $q + 1$ elements. 0 to q elements and 1 to $q - 1$ elements. Thus $|GL(2, F)/C| = (q + 1)q(q - 1)$ and since C contains $q - 1$ elements, $|GL(2, F)| = (q + 1)q(q - 1)^2$. Since $V_2(F)$ contains q^2 vectors, $GL(2, F)$ has order $(q^2 - 1)(q^2 - q)$.

24 It is an enlargement since three 1-dimensional subspaces are fixed. Compare qns 5 and 15.9.

25 $(1243): x \mapsto 2x$. $(0\infty)(13)(24): x \mapsto 3/x$.

	4		
∞	2	0	3
	1		

Now use qns 5.10 and 17.23.

26 Qn 13.16 shows that each element in the centre of $SL(2, F)$ is in the centre of $GL(2, F)$.

27 The multiplicative group of F contains $q - 1$ elements, an odd number, so by Lagrange's theorem every element of this group has odd order. $\begin{pmatrix} a & 0 \\ 0 & a \end{pmatrix}$ is in the centre of $SL(2, F) \Leftrightarrow a^2 = 1$.

29

F	$|GL(2, F)|$	$|SL(2, F)|$	$|C|$	$|C_S|$	$|GL(2, F)/C|$	$|SL(2, F)/C_S|$
$\mathbb{Z}_2$	6	6	1	1	6	6
$\mathbb{Z}_3$	48	24	2	2	24	12
F_4	180	60	3	1	60	60
$\mathbb{Z}_5$	480	120	4	2	120	60

31 If $A\pi \in S\pi$ then for some $B \in S$, $A\pi = B\pi$, so $AB^{-1}\pi = I\pi$ and AB^{-1} is a scalar matrix, $\begin{pmatrix} a & 0 \\ 0 & a \end{pmatrix}$ say. Thus det $AB^{-1} = a^2$, so det $A = a^2$.

32 $\begin{pmatrix} a/r & c/r \\ b/r & d/r \end{pmatrix}$.

33 From qns 31 and 32,

$$x \mapsto \frac{ax + b}{cx + d}$$

is in $S\pi$ if and only if $ad - bc$ is a square. So π is a homomorphism of $SL(2, F)$ onto the subgroup of $LF(F)$ with square determinants. The kernel of this homomorphism is C_S, and the fundamental theorem gives the result.

34 Both $(012): x \mapsto x + 1$ and $(0\infty)(12): x \mapsto 2/x$ have determinant 1. An element of order 2 and an element of order 3 generate A_4 by qn 17.40. Result follows from qns 29 and 33.

35 $(01234): x \mapsto x + 1$ and $(01\infty)(243): x \mapsto 4/(x + 4)$ each have determinant 1.

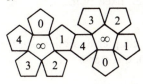

36 In $\mathbb{C}$ every element is a square, so $PGL(2, \mathbb{C}) \cong PSL(2, \mathbb{C})$.

19

Quaternions and rotations

In this chapter we first extend the algebra of matrices by defining matrix addition and establishing the distributive laws. We then use matrices to define the quaternions and develop some of their special algebraic properties. Finally we show how the group of inner automorphisms of the quaternions is isomorphic to the group of rotations of 3-dimensional space with a given fixed point.

Concurrent reading: Birkoff and MacLane, chapter 8, section 10; Rees, pp. 42–44; Curtis, chapter 5; Coxeter (1974), chapter 6.

Addition of matrices

In the first two questions we develop the algebra of matrices in a quite general context. We define addition on matrices of the same shape in a way which coincides with componentwise addition, and we show that this addition is compatible with our definition of matrix multiplication in the sense that both distributive laws hold.

1 If $\mathbf{v} \mapsto \mathbf{v}A$ and $\mathbf{v} \mapsto \mathbf{v}B$ are both linear transformations $V_n(F) \to V_m(F)$, prove that $\mathbf{v} \mapsto \mathbf{v}A + \mathbf{v}B$ is a linear transformation.
This establishes the existence of a matrix C such that $\mathbf{v}A + \mathbf{v}B = \mathbf{v}C$ for all vectors $\mathbf{v}$. We now *define* $A + B = C$ and this operation is called *addition of matrices*.
By considering $\mathbf{v} = (1, 0, \ldots, 0), (0, 1, \ldots, 0)$, etc., show that C is formed from A and B by componentwise addition.

2 If A, B and M are $n \times n$ matrices over the same field, use the definition of $A + B$ to prove that
$$M(A + B) = MA + MB.$$
In addition, by making an appeal to the linearity of $\mathbf{v} \mapsto \mathbf{v}M$,

prove that

$(A + B)M = AM + BM.$

Algebra of quaternions

We now study the set of matrices which we considered in the last question of the last chapter, and in qns 3–8 obtain the basic algebraic properties of this set.

3 We will call any complex 2×2 matrix of the form $\begin{pmatrix} z & w \\ -\bar{w} & \bar{z} \end{pmatrix}$ a *quaternion*, and we write every quaternion

$$\begin{pmatrix} a + ib & c + id \\ -c + id & a - ib \end{pmatrix} = a\begin{pmatrix} 1 & 0 \\ 0 & 1 \end{pmatrix} + b\begin{pmatrix} i & 0 \\ 0 & -i \end{pmatrix} + c\begin{pmatrix} 0 & 1 \\ -1 & 0 \end{pmatrix}$$
$$+ d\begin{pmatrix} 0 & i \\ i & 0 \end{pmatrix}$$

where a, b, c and d are real numbers, using the notation of scalar multiplication of matrices, namely $\begin{pmatrix} kp & kq \\ kr & ks \end{pmatrix} = k\begin{pmatrix} p & q \\ r & s \end{pmatrix}$.

Let $\mathbf{i} = \begin{pmatrix} i & 0 \\ 0 & -i \end{pmatrix}$, $\mathbf{j} = \begin{pmatrix} 0 & 1 \\ -1 & 0 \end{pmatrix}$, $\mathbf{k} = \begin{pmatrix} 0 & i \\ i & 0 \end{pmatrix}$.

Prove that $\mathbf{i}^2 = \mathbf{j}^2 = \mathbf{k}^2 = -I$, and that $\mathbf{ij} = \mathbf{k}$, $\mathbf{ji} = -\mathbf{k}$, $\mathbf{jk} = \mathbf{i}$, $\mathbf{kj} = -\mathbf{i}$, $\mathbf{ki} = \mathbf{j}$, $\mathbf{ik} = -\mathbf{j}$, to establish that the set of eight matrices $\{\pm I, \pm\mathbf{i}, \pm\mathbf{j}, \pm\mathbf{k}\}$ forms a group under matrix multiplication.

We use the notation of this question throughout the rest of the chapter.

4 Let α denote the function with domain the quaternions and codomain $V_4(\mathbb{R})$ given by

$$\alpha: \begin{pmatrix} a + ib & c + id \\ -c + id & a - ib \end{pmatrix} \mapsto (a, b, c, d).$$

Prove that α is a bijection, and that under α the structure of matrix addition corresponds with vector addition, and the structure of the scalar multiplication of matrices corresponds to scalar multiplication of vectors.

It is the isomorphism α which justifies the description of the quaternions as a 4-dimensional vector space over the real numbers.

5 If $A = aI + b\mathbf{i} + c\mathbf{j} + d\mathbf{k}$, where a, b, c, and d are real numbers, we define the *conjugate quaternion*

$$\bar{A} = aI - b\mathbf{i} - c\mathbf{j} - d\mathbf{k}.$$

Use qns 2 and 3 to show that

$$A\bar{A} = (a^2 + b^2 + c^2 + d^2)I$$

$$= (\det A)I.$$

The quaternion A given above is said to be a *real quaternion* if $b = c = d = 0$. Check that the real quaternions commute, under multiplication, with all quaternions.

6 Use qns 2 and 3 to show that the set of quaternions is closed under matrix multiplication. Use qn 5 to show that every nonzero quaternion has a multiplicative inverse which is a quaternion. Which of the axioms for a field can we now claim are valid for the quaternions under the operations of matrix addition and matrix multiplication?

7 Justify each of the equalities $(AB)(\overline{AB}) = (\det AB)I$ $= (\det A \cdot \det B)I = (A\bar{A})(B\bar{B}) = A(B\bar{B})\bar{A}$, and deduce that $\overline{AB} = \bar{B} \cdot \bar{A}$. Check that $\overline{A + B} = \bar{A} + \bar{B}$.

8 With the notation of qn 5 show that $Ai = iA$ implies $c = d = 0$ and that $Aj = jA$ implies $b = d = 0$. Deduce that the real quaternions form the centre of the multiplicative group of quaternions.

The transformation $X \mapsto R^{-1}XR$

We now examine the function $X \mapsto R^{-1}XR$ of the quaternions and see how it can be thought of as an isometry of real 3-dimensional space.

9 Let R be any nonzero quaternion and let ϱ denote the function with domain the quaternions given by $\varrho\colon X \mapsto R^{-1}XR$. Use qn 2 to prove that $R^{-1}(X + Y)R = R^{-1}XR + R^{-1}YR$. Explain why $R^{-1}(rX)R = r(R^{-1}XR)$ for any real number r. Then with the help of the bijection α of qn 4 explain why $X\alpha \mapsto (R^{-1}XR)\alpha$ is a linear transformation of $V_4(\mathbb{R})$.

10 Why must the linear transformation $X\alpha \mapsto (R^{-1}XR)\alpha$ be a bijection? Check that ϱ fixes every real quaternion.

11 Show that the characteristic equation of the matrix A of qn 5 is $\lambda^2 - 2a\lambda + \det A = 0$, and that $\det(R^{-1}AR - \lambda I) = \det[R^{-1}(A - \lambda I)R] = \det(A - \lambda I)$, so that A and $R^{-1}AR$ have the same characteristic equation.

The matrix A is said to be a *pure quaternion* when $a = 0$. Prove that ϱ acts as a bijection on the set of pure quaternions.

12 Let β denote the function with domain the pure quaternions and codomain $V_3(\mathbb{R})$ given by

$$\beta\colon b\mathbf{i} + c\mathbf{j} + d\mathbf{k} \mapsto (b, c, d).$$

Is β a bijection? Does β preserve the structure of vector addition and scalar multiplication? Deduce that $\beta^{-1}\varrho\beta$ is a nonsingular linear transformation of $V_3(\mathbb{R})$.

13 For any quaternions X and R with $R \neq 0$, prove that $(R^{-1}XR)(\overline{R^{-1}XR}) = X\bar{X}$. If X is a pure quaternion and the distance from the origin to $X\beta$ is x, prove that $X\bar{X} = x^2I$. Deduce that $X\beta \mapsto (R^{-1}XR)\beta$ preserves distances from the origin. By associating the distance between $X\beta$ and $Y\beta$ with $(X - Y)(\overline{X - Y})$, prove that $X\beta \mapsto (R^{-1}XR)\beta$ is an isometry of $V_3(\mathbb{R})$ fixing the origin.

The remaining questions in this chapter pinpoint the precise relationship between the quaternion R and the rotation $X\beta \mapsto (R^{-1}XR)\beta$ induced on $V_3(\mathbb{R})$ by $X \mapsto R^{-1}XR$.

14 If $R = aI + bU$ where a and b are real numbers and U is a pure quaternion, prove that $RU = UR$ and deduce that every quaternion rU, where r is real, is fixed by $X \mapsto R^{-1}XR$. This, in effect, identifies the axis of the rotation as $Sp(U\beta)$.

15 If $R^{-1}XR = Q^{-1}XQ$ for all quaternions X, deduce from qn 8 that QR^{-1} is a real quaternion and conversely. Deduce that for any $R \neq 0$ we may find a Q such that $R^{-1}XR = Q^{-1}XQ$ and $Q\bar{Q} = I$, a so-called *unit* quaternion.
Suppose now that $Q = aI + b\mathbf{i} + c\mathbf{j} + d\mathbf{k}$
and $a^2 + b^2 + c^2 + d^2 = 1$. If $a^2 = \cos^2\frac{1}{2}\theta$, then
$b^2 + c^2 + d^2 = \sin^2\frac{1}{2}\theta$. Deduce that Q may be written in the form
$\cos\frac{1}{2}\theta I + \sin\frac{1}{2}\theta U$,
where U is a pure quaternion and $U\bar{U} = I$. (If U is a unit pure quaternion, $U\beta$ is a point on the unit sphere, centre the origin.)

16 If X and Y are pure quaternions, we define $X \cdot Y = (X\beta) \cdot (Y\beta)$ with β as in qn 12 and the scalar product defined as in qn 14.3. We also define $X \times Y$ as $(X\beta \times Y\beta)\beta^{-1}$ as in qn 14.5. Prove that
$XY = (-X \cdot Y)I + X \times Y$.
If U is a pure quaternion with $U\bar{U} = I$ and X is a pure quaternion such that $X \cdot U = 0$ (so that, from the origin, $X\beta$ is in a direction at right angles to $U\beta$), prove that
$XU = X \times U = -UX = \bar{U}X = U^{-1}X$, so $UXU = X$.

17 Let $Q = \cos\frac{1}{2}\theta I + \sin\frac{1}{2}\theta U$ where U is a pure quaternion with $U\bar{U} = I$. If $X \cdot U = 0$, prove that $Q^{-1}XQ = \cos\theta X + \sin\theta X \times U$. (Remember $Q^{-1} = \cos\frac{1}{2}\theta I - \sin\frac{1}{2}\theta U$.) Check that $(Q^{-1}XQ) \cdot U = 0$ so that both $X\beta$ and $(Q^{-1}XQ)\beta$ lie in the plane through the origin perpendicular to the direction $OU\beta$. Use the

fact that from the origin the directions of $U\beta$, $X\beta$ and $(X \times U)\beta$ are mutually perpendicular to show that the transformation $X\beta \mapsto (Q^{-1}XQ)\beta$ acts as a rotation on the plane.

Since every pure quaternion Y corresponds to a point $Y\beta$ which is the sum of an $X\beta$ and a scalar multiple of $U\beta$, prove that the whole space is rotated about the same axis through an angle θ, by using qn 9.

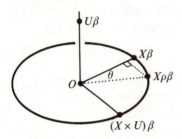

18 What is the kernel of the homomorphism $R \mapsto [X \mapsto R^{-1}XR]$? Compare this result with qn 18.37.

Summary

Theorem If A and B are $n \times m$ matrices over the same field F,
qn 1 there is a matrix C such that $\mathbf{v}A + \mathbf{v}B = \mathbf{v}C$ for all
  $\mathbf{v} \in V_n(F)$, and we define $A + B = C$.

Theorem Matrix multiplication is *distributive* over matrix addition
qn 2 both on the left and on the right.

Definition Complex matrices of the form
qn 3 $\begin{pmatrix} z & w \\ -\bar{w} & \bar{z} \end{pmatrix}$

  are called *quaternions*.

Theorem Under the operations of matrix addition and matrix
qns 2, 5, multiplication, quaternions satisfy all the axioms for a
6 field except for the commutativity of multiplication.

Theorem The quaternions form a 4-dimensional vector space over
qn 4 the real numbers.

Theorem $X \mapsto R^{-1}XR$ gives a linear transformation of the
qn 9 quaternions, and is a bijection.

Definition A quaternion of the form
qn 11 $b\begin{pmatrix} i & 0 \\ 0 & -i \end{pmatrix} + c\begin{pmatrix} 0 & 1 \\ -1 & 0 \end{pmatrix} + d\begin{pmatrix} 0 & i \\ i & 0 \end{pmatrix}$ is called a *pure*
  quaternion.

Theorem The pure quaternions form a 3-dimensional vector space
qn 12 over the real numbers.

Theorem $X \mapsto R^{-1}XR$ maps the pure quaternions onto
qns 11,   themselves and so acts as a linear transformation and in
12, 13   fact an isometry on this 3-dimensional space.
Theorem   If $R = r(\cos\frac{1}{2}\theta I + \sin\frac{1}{2}\theta U)$ where U is a pure
qns 14–17 quaternion of unit length, then $X \mapsto R^{-1}XR$ is a
           rotation with axis $Sp(U)$ through an angle θ.

Historical note

W. R. Hamilton tried to extend the way in which the complex
numbers describe the points of the real plane to a system of hyper-
complex numbers which would describe the points of real
3-dimensional space. There is of course no difficulty in defining
addition and scalar multiplication on the points of real 3-dimensional
space as on $V_3(\mathbb{R})$. The difficulty lies in defining a suitable
multiplication. Having failed in the task he set himself (which was
indeed impossible) he succeeded in 1843 in constructing an associative
multiplication on the points of $V_4(\mathbb{R})$. The ordered quadruples of real
numbers, with the operations of addition and multiplication which he
gave them, he called quaternions. They satisfy all the axioms for a field
except that multiplication is not commutative. Hamilton used the
quaternions to study the geometry of 3-dimensional space.

The development of quaternions from matrices which we used in
this chapter is due to A. Cayley (1858).

Answers to chapter 19

1 $(v_1A + v_1B) + (v_2A + v_2B) = (v_1 + v_2)A + (v_1 + v_2)B$ and
$k(vA + vB) = (kv)A + (kv)B$.

2 $[vM](A + B) = vMA + vMB$. $[v(A + B)]M = vAM + vBM$.

4 The first row of the matrix is uniquely determined by (a, b, c, d).

5 The real quaternions are precisely the real scalar matrices.

6 If $A = aI + bi + cj + dk$ then $(1/\det A)\bar{A}$ is its multiplicative
inverse unless $\det A = 0$. But $\det A = a^2 + b^2 + c^2 + d^2$, and $\det A = 0$
implies $a = b = c = d = 0$. All field axioms hold except that the multiplicative group is not abelian.

7 $(A\bar{A})(B\bar{B}) = A(B\bar{B})\bar{A}$ since $B\bar{B}$ is real.

9 Refer to the definition of a linear transformation in qn 12.2.

10 ϱ is an inner automorphism of the multiplicative group.

11 A quaternion is pure precisely when its characteristic equation has the form
$\lambda^2 + \det A = 0$.

13 $(R^{-1}XR)(\overline{R^{-1}XR}) = R^{-1}XR\bar{R}\bar{X}R^{-1}$
$\qquad = R^{-1}X\bar{X}R^{-1}(R\bar{R})$ since $R\bar{R}$ is real
$\qquad = (X\bar{X})R^{-1}R^{-1}(R\bar{R})$ since $X\bar{X}$ is real
$\qquad = X\bar{X}R^{-1}(R\bar{R})R^{-1}$.

A pure quaternion X has the form $bi + cj + dk$, and $X\beta = (b, c, d)$.
Distance of $X\beta$ from origin is $\sqrt{(b^2 + c^2 + d^2)}$. The square of the
distance between $X\beta$ and $Y\beta$ is a^2 where $a^2I = (X - Y)(\overline{X - Y})$.

14 $RU = (aI + bU)U = aU + bU^2 = U(aI + bU) = UR$. So
$R^{-1}UR = R^{-1}RU = U$.

15 $R^{-1}XR = Q^{-1}XQ \Leftrightarrow QR^{-1}X = XQR^{-1}$ so QR^{-1} is real. Thus $Q = aR$
(say). Now take $a = \sqrt{(R\bar{R})}$ and $Q\bar{Q} = I$.

16 $XU = X \times U$ since $X \cdot U = 0$.
$\qquad = -U \times X$
$\qquad = -UX$. For a pure quaternion $-U = \bar{U}$. For a unit quaternion
$\bar{U} = U^{-1}$.

17 $Q^{-1}XQ = (\cos\tfrac{1}{2}\theta I - \sin\tfrac{1}{2}\theta U)X(\cos\tfrac{1}{2}\theta I + \sin\tfrac{1}{2}\theta U)$
$\qquad = (\cos\tfrac{1}{2}\theta X - \sin\tfrac{1}{2}\theta UX)(\cos\tfrac{1}{2}\theta I + \sin\tfrac{1}{2}\theta U)$
$\qquad = \cos^2\tfrac{1}{2}\theta X - \sin\tfrac{1}{2}\theta\cos\tfrac{1}{2}\theta UX + \cos\tfrac{1}{2}\theta\sin\tfrac{1}{2}\theta XU - \sin^2\tfrac{1}{2}\theta UXU$
$\qquad = (\cos^2\tfrac{1}{2}\theta - \sin^2\tfrac{1}{2}\theta)X + 2\sin\tfrac{1}{2}\theta\cos\tfrac{1}{2}\theta XU$
$\qquad = \cos\theta X + \sin\theta X \times U$.

$Y = X + aU$ so $Q^{-1}YQ = Q^{-1}XQ + aU$.

18 The real quaternions.

20

Affine groups

Nonsingular linear transformations transform lines to lines and preserve ratios along a line. Translations also have these properties. The group generated by these two kinds of transformation is called the affine group. In the case of $V_n(\mathbb{R})$ the affine group is the full group of transformations preserving collinearity and also the full group of transformations preserving convexity.

Concurrent reading: Birkhoff and MacLane, chapter 9, section 3, pp. 252–5.

1 Express the function of complex numbers

$$z \mapsto e^{i\theta}z$$

as a linear transformation of $V_2(\mathbb{R})$ by matching the complex numbers $x + iy$ with the real points (x, y).

2 If the complex number c is equal to $a + ib$ where a and b are real, express the rotation

$$z \mapsto e^{i\theta}z + c$$

as a function of the vector space $V_2(\mathbb{R})$ onto itself.

Translation group

3 For any field F, a function of $V_2(F)$ of the form

$$(x, y) \mapsto (x, y) + (a, b)$$

is called a *translation*. Make a list of the four translations of $V_2(\mathbb{Z}_2)$ and of the nine translations of $V_2(\mathbb{Z}_3)$. How many translations has the vector space $V_2(\mathbb{Z}_p)$?

4 A function of the vector space $V_n(F)$ of the form

$$\mathbf{v} \mapsto \mathbf{v} + \mathbf{c}$$

is called a *translation* of the space. How many translations has the vector space $V_3(\mathbb{Z}_p)$? Must the translations of a vector space form a group? See 6.57.

Some particular affine transformations

5 For what real 2×2 matrix A and for what real 2-dimensional vector **c** is the transformation

$$\mathbf{v} \mapsto \mathbf{v}A + \mathbf{c}$$

of the vector space $V_2(\mathbb{R})$ equivalent to the mapping

$$(x, y) \mapsto (x + 2y + 3, 4x + 5y + 6)?$$

6 What conditions must be imposed on the 2×2 matrix A to ensure that the transformation of $V_2(\mathbb{R})$ given by

$$\mathbf{v} \mapsto \mathbf{v}A + \mathbf{c}$$

is a bijection?

7 Exhibit the transformations of the vector space $V_2(\mathbb{Z}_2)$ given by

$$(x, y) \mapsto (x, y)\begin{pmatrix} 1 & 1 \\ 0 & 1 \end{pmatrix} + (1, 0) \text{ and}$$

$$(x, y) \mapsto (x, y)\begin{pmatrix} 1 & 0 \\ 1 & 1 \end{pmatrix} + (1, 0)$$

as permutations of the four vectors in this space. Use qn 17.23 to prove that these transformations generate a group isomorphic to S_4.

Affine group

8 Let $\alpha : \mathbf{v} \mapsto \mathbf{v}A + \mathbf{c}$ and $\beta : \mathbf{v} \mapsto \mathbf{v}B + \mathbf{d}$ be transformations of the vector space $V_n(F)$, where A and B are $n \times n$ nonsingular matrices over F and **c** and **d** are vectors in this space.
  (i) If $\alpha = \beta$, by considering the image of **0**, prove that $\mathbf{c} = \mathbf{d}$ and deduce that $A = B$.
  (ii) Evaluate the product $\alpha\beta$ and check that it has the same form as α.
  (iii) Find β when $\alpha\beta$ is the identity.
  (iv) Deduce that the transformations of this type form a group.
  The group here is called the *affine group* $AG(n, F)$ and the transformations of which it consists are called *affine transformations*.

9 Compare the group $AG(1, \mathbb{Z}_3)$ with S_3.

10 Compare the group $AG(1, F_4)$, using qn 11.8, with A_4.

11 If $\alpha : \mathbf{v} \mapsto \mathbf{v}A + \mathbf{c}$ and $\tau : \mathbf{v} \mapsto \mathbf{v} + \mathbf{a}$ are affine transformations of $V_n(F)$, find $\alpha^{-1}\tau\alpha$, and deduce that the translation group T of a vector

space is a normal subgroup of the affine group. What is the quotient group $AG(n, F)/T$?

Lines in a vector space

12 If $\mathbf{0} = (0, 0)$ and $\mathbf{a} = (a, b) \neq \mathbf{0}$ are vectors in $V_2(\mathbb{R})$,
   (a) (i) describe the set of points $\{k\mathbf{a}|k \in \mathbb{R}\}$ geometrically;
      (ii) identify the subsets of this set for which $0 \leqslant k \leqslant 1$, $1 \leqslant k$ and $k \leqslant 0$ respectively.
   (b) (i) For a constant vector $\mathbf{c} \neq \mathbf{0}$, describe the set $\{k\mathbf{a} + \mathbf{c}|k \in \mathbb{R}\}$ geometrically;
      (ii) identify the subsets for which $0 \leqslant k \leqslant 1$, $1 \leqslant k$ and $k \leqslant 0$ respectively.

For any vector space, the 1-dimensional subspaces and their cosets in the additive group of vectors are called the *lines* of the space.

13 If in $V_n(F)$, $\mathbf{b} = \mathbf{a} + \mathbf{c}$, express the vector $k\mathbf{a} + \mathbf{c}$ as a linear combination of $\mathbf{c}$ and $\mathbf{b}$. If $\mathbf{u} = k\mathbf{a} + \mathbf{c}$, we say that $\mathbf{cu}:\mathbf{cb}$ are in the ratio $k:1$.

14 In the vector space $V_n(F)$ determine the set of points on the line through $\mathbf{u}$ and $\mathbf{v}$, where $\mathbf{u}$ and $\mathbf{v}$ are distinct points of the vector space.

Invariants of the affine group

15 If, under a translation τ of a vector space $V_n(F)$, $\mathbf{u}\tau = \mathbf{u}'$ and $\mathbf{v}\tau = \mathbf{v}'$, prove that $[(1 - k)\mathbf{u} + k\mathbf{v}]\tau = (1 - k)\mathbf{u}' + k\mathbf{v}'$. What does this imply about the images of lines of a vector space under a translation?

16 If, under a linear transformation α of a vector space $V_n(F)$, $\mathbf{u}\alpha = \mathbf{u}'$ and $\mathbf{v}\alpha = \mathbf{v}'$, prove that
$[(1 - k)\mathbf{u} + k\mathbf{v}]\alpha = (1 - k)\mathbf{u}' + k\mathbf{v}'$.

17 Let α be a line- and ratio-preserving transformation of the vector space $V_n(F)$ onto itself, in the sense that for any two distinct points of the space $\mathbf{u}$ and $\mathbf{v}$
$[(1 - k)\mathbf{u} + k\mathbf{v}]\alpha = (1 - k)(\mathbf{u}\alpha) + k(\mathbf{v}\alpha)$.
If τ is a translation of the space such that $\mathbf{0}\alpha = \mathbf{0}\tau$, does $\alpha\tau^{-1}$
  (i) fix $\mathbf{0}$,
  (ii) preserve lines and ratios?

18 If α is a line and ratio-preserving transformation of $V_n(F)$ which fixes $\mathbf{0}$, prove that for any vectors $\mathbf{u}$ and $\mathbf{v}$ in the space
  (i) $\alpha: k\mathbf{v} \mapsto k(\mathbf{v}\alpha)$,

(ii) for $k \neq 0, 1$, let $\mathbf{w} = (1 - k)^{-1}\mathbf{u}$ and $\mathbf{t} = k^{-1}\mathbf{v}$, find $\mathbf{w}\alpha$ and $\mathbf{t}\alpha$, examine $[(1 - k)\mathbf{w} + k\mathbf{t}]\alpha$ and prove that $\alpha : \mathbf{u} + \mathbf{v} \mapsto \mathbf{u}\alpha + \mathbf{v}\alpha$,
(iii) prove that α is a linear transformation.

19 Use qns 17 and 18 to show that every line- and ratio-preserving transformation of $V_n(F)$ is an affine transformation. Use qns 15 and 16 to show that every affine transformation of $V_n(F)$ preserves lines and ratios.

When a field F admits a nontrivial automorphism, the vector-space $V_n(F)$ admits line-preserving transformations which are not ratio-preserving.

Order of the affine group

20 Use qn 8(i) to prove that if F is a finite field, then $|AG(n, F)| = |GL(n, F)| \cdot |F|^n$.

Summary

Definition   For any vector space $V_n(F)$ the transformations of the
qn 4   form $\mathbf{v} \mapsto \mathbf{v} + \mathbf{a}$ are called *translations*.

Theorem   The translations of a vector space form a group iso-
qn 4   morphic to the group of vectors under addition.

Definition   If A is a matrix in $GL(n, F)$ then every transformation
qn 8   of $V_n(F)$ of the form

$$\mathbf{v} \mapsto \mathbf{v}A + \mathbf{c}$$

is called an *affine transformation*.

Theorem   The affine transformations of a vector space $V_n(F)$ form
qn 8   a group.

This group is called the affine group $AG(n, F)$.

Definition   The *lines* of $V_n(F)$ are the cosets of the 1-dimensional
qn 12   subspaces.

Definition   If $\mathbf{w} = (1 - k)\mathbf{u} + k\mathbf{v}$, then $\mathbf{uw} : \mathbf{uv}$ is in the ratio $k : 1$.
qn 13

Theorem   If $\mathbf{u}$ and $\mathbf{v}$ are distinct vectors of $V_n(F)$, then the unique
qn 14   line containing $\mathbf{u}$ and $\mathbf{v}$ is the set
$$\{(1 - k)\mathbf{u} + k\mathbf{v}|k \in F\}.$$

Theorem   All affine transformations preserve lines and ratios.
qn 19

Theorem   All line- and ratio-preserving transformations of $V_n(F)$
qn 19   are affine transformations.

Historical note

As long ago as 1762 E. Waring considered transformations of the plane of the type

$$(x, y) \mapsto \left(\frac{ax + by + c}{px + qy + r}, \frac{ex + fy + h}{px + qy + r} \right)$$

and claimed that they transformed curves into curves of the same degree. This approach developed into the study of projective transformations with homogeneous coordinates in the first half of the nineteenth century. Under a projective transformation any conic may be transformed to any other conic. The affine group is the subgroup of the projective group fixing the line at infinity, and affine transformations transform ellipses into ellipses, parabolae into parabolae and hyperbolae into hyperbolae, and as such was considered by J. Plücker (1835) and A. F. Möbius. Recognition of the importance of this group in its own right as the group of all line-preserving transformations of the real plane or real space dates from F. Klein's *Erlanger Programme* (1872). E. Galois (1832) identified the affine group over $V_n(\mathbb{Z}_p)$. C. Jordan in his *Traité des Substitutions* (1870) defined the translation group T of the vector space $V_n(\mathbb{Z}_p)$. He then considered the full set S of functions on $V_n(\mathbb{Z}_p)$ such that $ST = TS$ and showed that every element of S must be the product of a linear transformation and a translation.

Answers to chapter 20

1 $(\cos \theta + i \sin \theta)(x + iy) = x \cos \theta - y \sin \theta + i(x \sin \theta + y \cos \theta)$.

$(x, y) \mapsto (x, y) \begin{pmatrix} \cos \theta & \sin \theta \\ -\sin \theta & \cos \theta \end{pmatrix}$.

2 $(x, y) \mapsto (x \cos \theta - y \sin \theta + a, x \sin \theta + y \cos \theta + b)$.

3 In $V_2(\mathbb{Z}_2)$, $(a, b) = (0, 0), (1, 0), (0, 1)$ or $(1, 1)$. $V_2(\mathbb{Z}_p)$ admits p^2 translations.

4 $V_3(\mathbb{Z}_p)$ admits p^3 translations. From qn 6.57 the translations form a group isomorphic to the group of vectors.

5 $A = \begin{pmatrix} 1 & 4 \\ 2 & 5 \end{pmatrix}$, $\mathbf{c} = (3, 6)$.

6 A must be nonsingular.

7 $(0, 0) \mapsto (1, 0) \mapsto (0, 1) \mapsto (1, 1) \mapsto (0, 0)$; a 4-cycle. $(0, 0) \leftrightarrow (1, 0)$; a transposition.

8 (i) If $\tau : \mathbf{v} \mapsto \mathbf{v} + \mathbf{c}$, $\alpha = \beta \Rightarrow \alpha\tau^{-1} = \beta\tau^{-1} \Rightarrow \alpha\tau^{-1}(\beta\tau^{-1})^{-1} = 1$ so $AB^{-1} = I$.
(ii) $\alpha\beta : \mathbf{v} \mapsto \mathbf{v}AB + \mathbf{c}B + \mathbf{d}$. (iii) $\alpha^{-1} : \mathbf{v} \mapsto \mathbf{v}A^{-1} - \mathbf{c}A^{-1}$.

9 $(0); x \mapsto x, (012): x \mapsto x + 1, (021): x \mapsto x + 2, (12): x \mapsto 2x$,
$(01): x \mapsto 2x + 1, (02): x \mapsto 2x + 2$.

10 $(10)(ab): x \mapsto x + 1, (01b): x \mapsto ax + 1$ for example.

11 $\alpha^{-1}\tau\alpha : \mathbf{v} \mapsto \mathbf{v} + \mathbf{a}A$. $AG(n, F)/T \cong GL(n, F)$.

12 (a) (i) $Sp(\mathbf{a})$ is the line through the origin and (a, b). (ii) $0 \leqslant k \leqslant 1$ gives points between $\mathbf{0}$ and $\mathbf{a}$. $1 \leqslant k$ gives points beyond $\mathbf{a}$. $k \leqslant 0$ gives points beyond $\mathbf{0}$.
(b) (i) A line parallel to $Sp(\mathbf{a})$ through $\mathbf{c}$. (ii) $0 \leqslant k \leqslant 1$ gives points between $\mathbf{c}$ and $\mathbf{c} + \mathbf{a}$. $1 \leqslant k$ gives points beyond $\mathbf{c} + \mathbf{a}$. $k \leqslant 0$ gives points beyond $\mathbf{c}$.

13 $k\mathbf{a} + \mathbf{c} = k\mathbf{b} + (1 - k)\mathbf{c}$.

14 $\{k\mathbf{u} + (1 - k)\mathbf{v} | k \in F\}$.

15 If $\mathbf{u}\tau = \mathbf{u}'$ then $\tau : \mathbf{v} \mapsto \mathbf{v} + (\mathbf{u}' - \mathbf{u})$ for all $\mathbf{v}$, and $\mathbf{v}' - \mathbf{v} = \mathbf{u}' - \mathbf{u}$. So
$[(1 - k)\mathbf{u} + k\mathbf{v}]\tau = (1 - k)\mathbf{u} + k\mathbf{v} + (\mathbf{u}' - \mathbf{u})$
$= (1 - k)\mathbf{u} + k\mathbf{v} + (1 - k)(\mathbf{u}' - \mathbf{u}) + k(\mathbf{v}' - \mathbf{v})$.
Lines are mapped to lines under a translation.

18 (i) Put $\mathbf{u} = \mathbf{0}$. (iii) Implication of (i) and (ii).

20 Number of nonsingular matrices $= |GL(n, F)|$. Number of translations $=$ number of vectors.

21

Orthogonal groups

The first step in this chapter is to identify the subgroup of isometries in the general linear group, and this is in fact the full group of isometries stabilising a point. In the context of linear groups these are called orthogonal transformations. All orthogonal transformations have determinant $+1$ or -1.

The second step in this chapter is to show that in both two and three dimensions, every orthogonal transformation with determinant $+1$ is a rotation.

The third step, taken by considering an arbitrary isometry as an affine transformation, is to prove that every finite group of isometries stabilises a point and is thus isomorphic to a subgroup of the orthogonal group.

The final step is to classify finite groups of rotations in three dimensions as being isomorphic to either C_n, D_n, A_4, S_4 or A_5, and then to extend these groups to include opposite isometries.

Concurrent reading: Gardiner, chapter 4; Weyl, appendices.

The first eight questions of this chapter establish the equivalence of the following three statements in relation to either $V_2(\mathbb{R})$ or $V_3(\mathbb{R})$.

   (i) α is an isometry fixing the origin;
   (ii) α preserves scalar products, i.e. $\mathbf{u} \cdot \mathbf{v} = \mathbf{u}\alpha \cdot \mathbf{v}\alpha$;
   (iii) α is a linear transformation $\alpha : \mathbf{v} \mapsto \mathbf{v}A$ and $AA^T = I$.

The transformations satisfying these conditions will be called orthogonal transformations, and their matrices, orthogonal matrices.

  1 In the group of isometries of the Euclidean plane, identify the stabiliser of the origin (qn 3.19).

  2 By identifying the complex number $z = x + \mathrm{i}\,y$ with the vector (x, y)

express both $z \mapsto e^{i\theta}z$ and $z \mapsto e^{i\theta}\bar{z}$ as linear transformations of $V_2(\mathbb{R})$.

When the group of isometries fixing the origin is considered as a subgroup of $GL(2, \mathbb{R})$ it is called *the orthogonal group $O(2)$*.
For each of the matrices A that you have used to express these isometries as linear transformations, work out the product AA^T.

3 What are the determinants of the matrices in $O(2)$? Which elements of $O(2)$ form the kernel of the homomorphism $O(2) \to \mathbb{R}$ given by $A \mapsto \det A$?

This subgroup is known as the *special orthogonal group $SO(2)$*. Describe the elements of this subgroup geometrically.

4 If $\mathbf{u} = (u_1, u_2)$ and $\mathbf{v} = (v_1, v_2)$ are vectors of $V_2(\mathbb{R})$, we define the scalar product $\mathbf{u}\cdot\mathbf{v} = u_1v_1 + u_2v_2$ by analogy with qn 14.3.
  (i) Prove that $\mathbf{u}\cdot\mathbf{u} = 0$ only when $\mathbf{u} = \mathbf{O}$.
  (ii) Prove that $\mathbf{u}\cdot\mathbf{v} = \mathbf{v}\cdot\mathbf{u}$.
  (iii) Prove that $\mathbf{u}\cdot(\mathbf{v} + \mathbf{w}) = \mathbf{u}\cdot\mathbf{v} + \mathbf{u}\cdot\mathbf{w}$.
  (iv) Prove that the distance between the two points
    $\mathbf{u}$ and $\mathbf{v}$ is $\sqrt{[(\mathbf{u} - \mathbf{v}) \cdot (\mathbf{u} - \mathbf{v})]}$.
  (v) If α is an isometry fixing the origin, by considering the distance from $\mathbf{O}$ to $\mathbf{u}$ prove that $\mathbf{u} \cdot \mathbf{u} = \mathbf{u}\alpha \cdot \mathbf{u}\alpha$, and by considering the distance from $\mathbf{u}$ to $\mathbf{v}$ prove that $\mathbf{u}\cdot\mathbf{v} = \mathbf{u}\alpha\cdot\mathbf{v}\alpha$.
  (vi) Check that all the results of this question hold for $V_3(\mathbb{R})$.

5 If α is a transformation of $V_2(\mathbb{R})$ or $V_3(\mathbb{R})$ such that $\mathbf{u}\cdot\mathbf{v} = \mathbf{u}\alpha\cdot\mathbf{v}\alpha$ for all vectors $\mathbf{u}$ and $\mathbf{v}$, prove that
  (i) $\mathbf{u}\cdot\mathbf{u} = \mathbf{u}\alpha\cdot\mathbf{u}\alpha$,
  (ii) $\mathbf{O}\alpha = \mathbf{O}$,
  (iii) α is an isometry fixing the origin.

Qns 4 and 5 together provide the proof that the isometries fixing the origin are precisely the scalar-product-preserving transformations.

6 If $\mathbf{v} \mapsto \mathbf{v}A$ is an isometry of $V_3(\mathbb{R})$ and the rows of the matrix A are the vectors $\mathbf{a}$, $\mathbf{b}$ and $\mathbf{c}$ respectively, use qn 4(v) to show that $\mathbf{a}\cdot\mathbf{a} = \mathbf{b}\cdot\mathbf{b} = \mathbf{c}\cdot\mathbf{c} = 1$ and $\mathbf{a}\cdot\mathbf{b} = \mathbf{b}\cdot\mathbf{c} = \mathbf{c}\cdot\mathbf{a} = 0$. Deduce that $AA^T = I$.

7 We now prove that scalar-product-preserving transformations are necessarily linear. Let α be a transformation of $V_2(\mathbb{R})$ such that $\mathbf{u}\cdot\mathbf{v} = \mathbf{u}\alpha\cdot\mathbf{v}\alpha$. From qn 5, α is an isometry of $V_2(\mathbb{R})$ and hence a bijection. Suppose $(1, 0)\alpha = (a_1, a_2)$ and that $(0, 1)\alpha = (b_1, b_2)$. By working out scalar products, prove that if $(x, y)\alpha = (x', y')$,

then $(x, y) = (x', y') \begin{pmatrix} a_1 & b_1 \\ a_2 & b_2 \end{pmatrix}$. This establishes that

α^{-1}:$(x', y') \mapsto (x, y)$ is a linear transformation. Now α and thus α^{-1} are bijections and so α^{-1} is a nonsingular linear transformation, and thus α is too. Check that the matrix for α^{-1} is the transpose of the matrix for α, so that for α:$\mathbf{v} \mapsto \mathbf{v}A$, $AA^{\mathrm{T}} = I$.
Check that the working here can be carried through in $V_3(\mathbb{R})$.

8 Let α:$\mathbf{v} \mapsto \mathbf{v}A$ be a linear transformation of either $V_2(\mathbb{R})$ or $V_3(\mathbb{R})$ for which $AA^{\mathrm{T}} = I$. Use the fact that $\mathbf{u}\cdot\mathbf{v} = \mathbf{u}\mathbf{v}^{\mathrm{T}}$ to check that α preserves scalar products.

Qns 6, 7 and 8 together provide the proof that the scalar-product-preserving transformations are precisely the linear transformations $\mathbf{v} \mapsto \mathbf{v}A$ for which $AA^{\mathrm{T}} = I$.

9 The isometries in $GL(3, \mathbb{R})$ are called *orthogonal transformations* and are said to form the *orthogonal group* $O(3)$. A matrix A for which $AA^{\mathrm{T}} = I$ is called an *orthogonal matrix*.
Must the orthogonal matrices form a group under matrix multiplication? Prove that the determinant of an orthogonal matrix is ± 1. By using the ideas of qn 3 show that the orthogonal matrices with determinant $+1$ form a subgroup of the orthogonal group. The subgroup of $O(3)$ of matrices with determinant $+1$ is called the *special orthogonal group* $SO(3)$.

10 Check that each of the following matrices is orthogonal and find the determinant in each case.

$$\begin{pmatrix} \cos\theta & \sin\theta & 0 \\ -\sin\theta & \cos\theta & 0 \\ 0 & 0 & 1 \end{pmatrix},$$

$$\begin{pmatrix} \cos\theta & \sin\theta & 0 \\ \sin\theta & -\cos\theta & 0 \\ 0 & 0 & 1 \end{pmatrix}, \quad \begin{pmatrix} \cos\theta & \sin\theta & 0 \\ -\sin\theta & \cos\theta & 0 \\ 0 & 0 & -1 \end{pmatrix}.$$

If A denotes the last of these matrices, check that $\mathbf{v} \mapsto \mathbf{v}A$ fixes only the origin and no other points when $\theta \neq 2n\pi$.

The special orthogonal group $SO(3)$

Through our work with complex numbers we already know that $SO(2)$ consists wholly of rotations, and all the remaining elements of $O(2)$ are reflections. We have just seen, through the last part of qn 10, that in three dimensions there are orthogonal transformations with determinant -1 which are not reflections. It is therefore somewhat surprising to find that every element of $SO(3)$ is a rotation, a fact that we establish in the next six questions.

11 If A is a matrix in $SO(3)$, justify each line of the following argument.
$$
\begin{aligned}
\det (A - I) &= \det (A - AA^{\mathrm{T}}) \\
&= \det A(I - A^{\mathrm{T}}) \\
&= \det A \cdot \det (I - A^{T}) \\
&= \det (I - A^{\mathrm{T}}) \\
&= \det (I - A) \\
&= (-1)^3 \det (A - I).
\end{aligned}
$$
Deduce finally that $\det (A - I) = 0$.

12 If A is a matrix in $SO(3)$, use qn 11 to show that A has an eigenvalue $+1$.

13 If $\mathbf{a}$ is a given nonzero vector of $V_3(\mathbb{R})$, describe geometrically the set of vectors $\{\mathbf{u} | \mathbf{u} \cdot \mathbf{a} = 0\}$.
If $\mathbf{a}$ is an eigenvector with eigenvalue 1 of the orthogonal transformation $\mathbf{v} \mapsto \mathbf{v}A$, use the preservation of scalar products under orthogonal transformations to find the value of $\mathbf{u}A \cdot \mathbf{a}$ when $\mathbf{u} \cdot \mathbf{a} = 0$. What can you deduce about the image of the plane through $\mathbf{O}$ perpendicular to the line $\mathbf{Oa}$ under the orthogonal transformation $\mathbf{v} \mapsto \mathbf{v}A$?

14 If $(0, 0, 1)$ is an eigenvector with eigenvalue 1 of the orthogonal transformation $\alpha: \mathbf{v} \mapsto \mathbf{v}A$, use qn 13 to show that α acts as an isometry on the xy-plane, and deduce the two possible forms of the matrix A. Distinguish which of these forms is possible if $A \in SO(3)$.

We know that every transformation in $SO(3)$ has an eigenvector with eigenvalue $+1$, and we have now shown that if this eigenvector happens to be $(0, 0, 1)$ the transformation is a rotation. We now generalise this result to show that it holds whatever the eigenvector may be.

15 We prove that $SO(3)$ is transitive on the points of the unit sphere, centre the origin. Let
$$
A = \begin{pmatrix} \cos \theta & \sin \theta & 0 \\ -\sin \theta & \cos \theta & 0 \\ 0 & 0 & 1 \end{pmatrix} \quad \text{and} \quad B = \begin{pmatrix} \cos \phi & 0 & -\sin \phi \\ 0 & 1 & 0 \\ \sin \phi & 0 & \cos \phi \end{pmatrix}.
$$
Verify that $A, B \in SO(3)$. Describe $\mathbf{v} \mapsto \mathbf{v}A$ and $\mathbf{v} \mapsto \mathbf{v}B$ geometrically. If (a_1, a_2, a_3) is a point distant one unit from the origin, determine suitable θ and ϕ so that under the transformation $\mathbf{v} \mapsto \mathbf{v}AB$ (a_1, a_2, a_3) has image $(0, 0, 1)$.

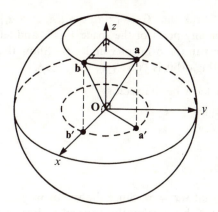

16 If α is an arbitrary transformation in $SO(3)$ with the eigenvector **a** with eigenvalue $+1$, and **a** is distant one unit from the origin, and if β is an element of $SO(3)$ for which $\mathbf{a}\beta = (0, 0, 1)$, must the transformation $\beta^{-1}\alpha\beta$ be in $SO(3)$, and what is its eigenvector? Use qn 14 to identify this transformation, and then qn 17.50 to identify α.

The orthogonal group $O(3)$

By finding a particular matrix which commutes with every element of the group and has determinant -1, we can use our knowledge of $SO(3)$ to describe the full orthogonal group.

17 Find a scalar matrix distinct from the identity in $O(3)$.

18 Use qn 10.11 to prove that the group $O(3)$ is isomorphic to the direct product of $SO(3)$ with $\{\pm\ I\}$.

Finite groups of isometries

We will prove that every finite group of isometries has a fixed point and is therefore isomorphic to a subgroup of the appropriate orthogonal group. In order to prove this we must first establish that every isometry is an affine transformation. We already have this result in two dimensions, but we need to prove it in three dimensions.

19 Let $\alpha : \mathbf{v} \mapsto \mathbf{v}A + \mathbf{c}$ be an affine transformation of $V_3(\mathbb{R})$. If α is an isometry, and we presume that the translation $\tau : \mathbf{v} \mapsto \mathbf{v} + \mathbf{c}$ is an isometry, prove that A is an orthogonal matrix.

20 If α is any isometry of $V_3(\mathbb{R})$ and τ is the translation such that $\mathbf{O}\alpha = \mathbf{O}\tau$, must $\alpha\tau^{-1}$ be an orthogonal transformation? Deduce that every isometry of $V_3(\mathbb{R})$ is an affine transformation.

21 Let G be a finite group of isometries of $V_3(\mathbb{R})$ of order n. Specifically

let $G = \{\alpha_1, \alpha_2, \ldots, \alpha_n\}$. Check that $G = \{\alpha_1\alpha_j, \alpha_2\alpha_j, \ldots, \alpha_n\alpha_j\}$ for any $\alpha_j \in G$. Now let $\mathbf{v}$ be any point of the space $V_3(\mathbb{R})$, and let $\mathbf{v}_i = \mathbf{v}\alpha_i$. What are the points in the orbit of $\mathbf{v}$ under G? Show that another way of describing this orbit is $\{\mathbf{v}_1\alpha_j, \mathbf{v}_2\alpha_j, \ldots, \mathbf{v}_n\alpha_j\}$.

22 With the notation of qn 21 let $\mathbf{w}$ be the vector

$$\frac{1}{n}(\mathbf{v}_1 + \mathbf{v}_2 + \ldots + \mathbf{v}_n)$$

and let

$$\alpha{:}\mathbf{v} \mapsto \mathbf{v}A + \mathbf{c}$$

be an element of G. Prove that $\mathbf{w}\alpha = \mathbf{w}$. Since the vector $\mathbf{w}$ depends on the whole group G but not on the particular element α, this means we have found a point which is stabilised by G.

23 By a suitable choice of the translation τ, show that any finite group of isometries, G, either of $V_2(\mathbb{R})$ or of $V_3(\mathbb{R})$ is isomorphic to a subgroup of $O(2)$ or $O(3)$ of the form $\tau G \tau^{-1}$.

24 If a finite subgroup of $SO(2)$ consists of rotations through angles θ_1, $\theta_2, \ldots, \theta_n$, where

$$0 = \theta_1 < \theta_2 < \ldots < \theta_n < 2\pi,$$

explain why a rotation through any multiple of θ_2 belongs to the group, and why, if a rotation other than by a multiple of θ_2 were to belong to the group, a contradiction would follow. (Argue as in qn 6.28.) Deduce that every finite subgroup of $SO(2)$ is cyclic.

25 If G is a finite subgroup of $O(2)$ containing at least one reflection, use qn 3 to show that the group G consists of an equal number of rotations and reflections.

26 If $\alpha{:}z \mapsto e^{i\theta}z$ and $\beta{:}z \mapsto e^{i\phi}\bar{z}$, evaluate $\beta\alpha\beta$ and deduce from qn 6.43 that a finite subgroup of $O(2)$ which contains at least one reflection is a dihedral group.

Following Weyl, the results of qns 24 and 26 have recently been called '*Leonardo's theorem*'.

Finite subgroups of $SO(3)$

Throughout qns 27–39, G denotes a finite subgroup of $SO(3)$.

27 Obtain a clean tennis ball or polystyrene sphere, and mark on it, in black, the six vertices of a prism with an equilateral triangle as cross-section. Consider the centre of the sphere as origin, and let G denote the largest subgroup of $SO(3)$ which is a symmetry group of the prism, that is, the full group of rotational symmetries of the prism. What is the order of G?. Check your answer with qn 9.18. For the proper subgroup of order 3, mark in green the points on the sphere which are stabilised by this subgroup. For the proper subgroups of G of order 2, mark in red the points on the sphere which are stabilised by one of these subgroups. Count the points which you have marked in green or red – these are called the *poles* of G – and determine the full orbit of each of these poles under G. Do each of the poles in one orbit have stabilisers of the same order?

28 On a fresh sphere, mark the vertices of a cube in black. Let G denote the rotational symmetry group of the cube. Mark in blue all those points on the sphere which are stabilised by a cyclic subgroup of G of order 4. Mark in green all those points of the sphere which are stabilised by a subgroup of G of order 3. Mark in red those points, not previously marked, which are stabilised by a subgroup of order 2. The points marked in red, green and blue are called the *poles* of G. Find the orbit of each pole of G under G. Do all the poles of an orbit have the same colour?

29 We call the blue points of qn 28 *4-poles*, the green points of qns 27 and 28 *3-poles* and the red points of qns 27 and 28 *2-poles*. In general, we consider a finite subgroup G of $SO(3)$ as acting on the unit sphere with centre at the origin, and call a point of the sphere an *m-pole* if it is on the axis of a rotation in G of order $m \geqslant 2$, always choosing the greatest possible m. Why is the length of the orbit to which an *m*-pole belongs $|G|/m$? Use qn 9.17.

30 Why must all the points in the orbit of an *m*-pole be *m*-poles? Use qn 9.17 or 17.50.

31 How many rotations in G, different from the identity, share the same *m*-pole?

32 If the numbers $(|G|/m)(m-1)$, obtained from each orbit of poles, are added together, why will every rotation in G, other than the identity, have been counted twice?

33 Writing $|G| = N$, deduce from qn 32 that

$$2(N - 1) = \sum_{\substack{\text{orbits} \\ \text{of poles}}} \frac{N(m - 1)}{m}$$

and hence

$$2 - 2/N = \sum_{\substack{\text{orbits} \\ \text{of poles}}} (1 - 1/m).$$

34 Explain why $1 \leqslant 2 - 2/N < 2$, and why $\frac{1}{2} \leqslant 1 - 1/m < 1$ in qn 33. Deduce that only one orbit of poles is impossible and that four or more orbits of poles are impossible.

35 Suppose there are two orbits of poles, one of m-poles and one of n-poles, so $2 - 2/N = (1 - 1/m) + (1 - 1/n)$. Deduce that $N/m + N/n = 2$ and so $m = n = N$. Since the rotations stabilising an N-pole exhaust the elements of G, there are just two poles and G is cyclic. Illustrate such poles for the rotational symmetries of a right regular pyramid.

36 Suppose there are three orbits of poles, one of l-poles, one of m-poles and one of n-poles, so
$2 - 2/N = (1 - 1/l) + (1 - 1/m) + (1 - 1/n)$.
Deduce that
$1 + 2/N = 1/l + 1/m + 1/n$
and that not all of l, m and n can be greater than or equal to 3.

37 In qn 36, take $2 = l \leqslant m \leqslant n$. If $m = 2$, prove that $n = \frac{1}{2}N$. Exhibit such poles for the group of rotational symmetries of a regular prism inscribed in a sphere. Thus $G \cong D_n$.

38 In qn 36, with $2 = l \leqslant m \leqslant n$, take $m = 3$ and prove
$$3 \leqslant n = \frac{6N}{N + 12} < 6.$$
If $n = 3$, prove that $N = 12$ and that the poles match those of the group of rotational symmetries of a regular tetrahedron inscribed in a sphere. If $n = 4$, prove that $N = 24$ and that the poles match those of the group of rotational symmetries of a cube inscribed in a sphere. If $n = 5$, prove that $N = 60$ and that the poles match those of the group of rotational symmetries of a regular icosahedron inscribed in a sphere. Check that $n \geqslant 6$ is impossible.

39 In qn 36, with $2 = l < 3 < m \leqslant n$, prove that
$1/l + 1/m + 1/n \leqslant 1 < 1 + 2/N$ which is impossible.

Finite subgroups of $O(3)$

40 If G is a finite subgroup of $O(3)$, use determinants to prove that *either* all the elements of G *or* half the elements of G belong to $SO(3)$.

4i A cuboid has eight vertices 1, 2, 3, 4, a, b, c, d, with 1234 and $abcd$ forming square faces and $1a$, $2b$, $3c$ and $4d$ being parallel edges. Write down the rotational symmetries of the cuboid as permutations of the vertices, and, after identifying a pair of generators for this group, α of order 4 and β of order 2, label the permutations e, α, α^2, α^3, β, $\beta\alpha$, $\beta\alpha^2$, $\beta\alpha^3$ respectively.
Write down the remaining eight symmetries of the cuboid as

permutations of the vertices, and attempt to describe these symmetries geometrically. If γ denotes the point symmetry $(1c)(2d)(3a)(4b)$, use qns 17 and 18 to prove that γ is in the centre of the full group of symmetries of the cuboid and that this group is isomorphic to $D_4 \times C_2$.

42 Name a group which is isomorphic to the full symmetry group of a cube.

43 A triangular prism has six vertices 1, 2, 3, a, b, c, with 123 and abc forming equilateral triangular faces and $1a$, $2b$ and $3c$ being equal and parallel edges. Write down the rotational symmetries of the prism as permutations of the vertices, and, after identifying a pair of generators for this group, α of order 3 and β of order 2, label the permutations e, α, α^2, β, $\beta\alpha$, $\beta\alpha^2$ respectively.

    Write down the remaining six symmetries of the prism as permutations of the vertices and attempt to describe them geometrically.

    The prism is inscribed in a sphere with centre O, and $11'$, $22'$, $33'$, aa', bb', cc' are diameters of this sphere. If γ denotes the point symmetry about O, which when restricted to the twelve named points on the sphere is $(11')(22')(33')(aa')(bb')(cc')$, and if δ denotes the rotational symmetry of the sphere through an angle $\pi/3$ which when restricted to the twelve named points on the sphere is $(1c'2a'3b')(1'c2'a3'b)$, verify that the six symmetries of the prism which are not rotations are $\delta\gamma$, $\delta^3\gamma$, $\delta^5\gamma$, $\beta\delta\gamma$, $\beta\delta^3\gamma$, $\beta\delta^5\gamma$. Is the full symmetry group of the prism a dihedral group? Is $\langle \delta, \beta \rangle$ a dihedral group?

44 If G is a subgroup of $O(3)$ and $G \cap SO(3) = H$, and if, moreover, G contains the point symmetry $\mathbf{v} \mapsto -\mathbf{v}$, prove that G is isomorphic to $H \times C_2$.

45 If G is a subgroup of $O(3)$ of order $2n$ and $G \cap SO(3) = \{\alpha_1, \alpha_2, \ldots, \alpha_n\}$ and if γ denotes the point symmetry $\mathbf{v} \mapsto -\mathbf{v}$, use qn 18 to show that the elements of G which are not in $SO(3)$ have the form $\beta_1\gamma$, $\beta_2\gamma$, \ldots, $\beta_n\gamma$, where the β_i are rotations. Deduce that either
$$\{\beta_1, \beta_2, \ldots, \beta_n\} = \{\alpha_1, \alpha_2, \ldots, \alpha_n\},$$
or else the $2n$ rotations α_i and β_i form a subgroup of $SO(3)$ of order $2n$.

Summary

Theorem   If α is a transformation of $V_2(\mathbb{R})$ or $V_3(\mathbb{R})$ to itself, then
qns 1–8   the three statements which follow are equivalent.

(i) α is an isometry fixing **O**;

(ii) α preserves scalar products;

(iii) α is a linear transformation $\mathbf{v} \mapsto \mathbf{v}A$ and $AA^{\mathrm{T}} = I$.

Definition
qns 2, 9
A transformation satisfying the conditions of the previous theorem is called an *orthogonal transformation* and a matrix A such that $AA^{\mathrm{T}} = I$ is called an *orthogonal matrix*.

Definition
qns 3, 9
The full group of orthogonal transformations in 2 (resp. 3) dimensions is denoted by $O(2)$ (resp. $O(3)$). The subgroup of elements with determinant 1 is denoted by $SO(2)$ (resp. $SO(3)$).

Theorem
qns 3,
11–16
Every element of $SO(2)$ and $SO(3)$ is a rotation.

Theorem
qns 18
The group $O(3)$ is isomorphic to the direct product of its subgroups $SO(3)$ and $\{\pm I\}$.

Theorem
qns 19–23
A finite group of isometries has a fixed point and is isomorphic to a subgroup of an orthogonal group.

Theorem
qns 24, 26
A finite subgroup of $SO(2)$ must be cyclic. Any other finite subgroup of $O(2)$ is dihedral.

Theorem
qns 27–39
A finite subgroup of $SO(3)$ is either cyclic, dihedral or the group of rotations of a regular solid.

Historical note

The orthogonal group was first studied as the group of transformations preserving the quadratic form $x^2 + y^2$ or $x^2 + y^2 + z^2$ by the number theorists of the eighteenth century. In C. Jordan's *Traité des Substitutions* (1870) the orthogonal group is named and defined as the group of linear transformations preserving $x^2 + y^2 + z^2 + \ldots$ and Jordan claims the details of the result $AA^{\mathrm{T}} = I$, but without using matrices. The possibility of defining orthogonal transformations in terms of the preservation of a scalar product (or more generally, of a symmetric bilinear form) followed the definition of the scalar product by J. W. Gibbs (1881).

The first proof of the classification of finite groups of rotations in 3-dimensional space is due to A. Bravais (1849) and this proof is given by H. S. M. Coxeter in his book *Regular Polytopes*. The proof we offer here stems from a discussion of Möbius transformations given by F. Klein in his *Lectures on the Icosahedron* (1884). The full symmetry groups of the regular polyhedra are indicated by F. Klein in that book and then left as an exercise for the reader.

Answers to chapter 21

1 $z \mapsto e^{i\theta}z$ and $z \mapsto e^{i\theta}\bar{z}$ for all real θ.

2 Matrix for rotation $\begin{pmatrix} \cos\theta & \sin\theta \\ -\sin\theta & \cos\theta \end{pmatrix}$ as in qn 20.1. Matrix for reflection $\begin{pmatrix} \cos\theta & \sin\theta \\ \sin\theta & -\cos\theta \end{pmatrix}$ found similarly. $AA^{\mathsf{T}} = I$.

3 Determinants are ± 1. The rotations form the kernel.

4 (iv) Distance from **u** to **v** is $\sqrt{[(u_1 - v_1)^2 + (u_2 - v_2)^2]}$. (v) Put $\mathbf{v} = \mathbf{O}$ in (iv) to get $\mathbf{u} \cdot \mathbf{u} = \mathbf{u}\alpha \cdot \mathbf{u}\alpha$. Now $(\mathbf{u} - \mathbf{v}) \cdot (\mathbf{u} - \mathbf{v}) = (\mathbf{u}\alpha - \mathbf{v}\alpha) \cdot (\mathbf{u}\alpha - \mathbf{v}\alpha)$ and we obtain $\mathbf{u} \cdot \mathbf{v} = \mathbf{u}\alpha \cdot \mathbf{v}\alpha$ using (iii), (ii) and $\mathbf{u} \cdot \mathbf{u} = \mathbf{u}\alpha \cdot \mathbf{u}\alpha$.

5 (iii) From $\mathbf{u} \cdot \mathbf{v} = \mathbf{u}\alpha \cdot \mathbf{v}\alpha$ we deduce $(\mathbf{u} - \mathbf{v}) \cdot (\mathbf{u} - \mathbf{v}) = (\mathbf{u}\alpha - \mathbf{v}\alpha) \cdot (\mathbf{u}\alpha - \mathbf{v}\alpha)$.

6 $(1, 0, 0)\alpha = \mathbf{a}$, so from qn 4(v), $\mathbf{a} \cdot \mathbf{a} = 1$. $(0, 1, 0)\alpha = \mathbf{b}$, so from qn 4 (v), $\mathbf{a} \cdot \mathbf{b} = 0$.
$$AA^{\mathsf{T}} = \begin{pmatrix} \mathbf{a} \cdot \mathbf{a} & \mathbf{a} \cdot \mathbf{b} & \mathbf{a} \cdot \mathbf{c} \\ \mathbf{b} \cdot \mathbf{a} & \mathbf{b} \cdot \mathbf{b} & \mathbf{b} \cdot \mathbf{c} \\ \mathbf{c} \cdot \mathbf{a} & \mathbf{c} \cdot \mathbf{b} & \mathbf{c} \cdot \mathbf{c} \end{pmatrix} = I.$$

7 $(1, 0) \cdot (x, y) = x = (a_1, a_2) \cdot (x', y')$. $(0, 1) \cdot (x, y) = y = (b_1, b_2) \cdot (x', y')$.

8 $\mathbf{u}A \cdot \mathbf{v}A = \mathbf{u}A(\mathbf{v}A)^{\mathsf{T}} = \mathbf{u}AA^{\mathsf{T}}\mathbf{v}^{\mathsf{T}} = \mathbf{u}\mathbf{v}^{\mathsf{T}} = \mathbf{u} \cdot \mathbf{v}$.

9 The orthogonal group consists of the isometries stabilising the origin. See qn 13.34. $\det AA^{\mathsf{T}} = 1$ and $\det A = \det A^{\mathsf{T}}$ so $\det A = \pm 1$. $A \mapsto \det A$ is a homomorphism. A is in the kernel if and only if $\det A = 1$.

10 $1, -1, -1$. $(x, y, z) = (x\cos\theta - y\sin\theta, x\sin\theta + y\cos\theta, -z) \Rightarrow$ $x = y = z = 0$ when $\theta \neq 2n\pi$.

12 $\det(A - I) = 0$ means $\lambda = 1$ is a solution of $\det(A - \lambda I) = 0$.

13 The plane through the origin perpendicular to $\mathbf{Oa}$. $\mathbf{u} \cdot \mathbf{a} = 0 \Rightarrow \mathbf{u}A \cdot \mathbf{a}A = 0$ $\Rightarrow \mathbf{u}A \cdot \mathbf{a} = 0$. So the plane through $\mathbf{O}$ perpendicular to $\mathbf{Oa}$ is fixed by $\mathbf{v} \mapsto \mathbf{v}A$.

14 A has one of the first two forms of qn 10. If $\det A = 1$ only the first is possible.

15 $\mathbf{v} \mapsto \mathbf{v}A$ is a rotation through θ about the z-axis. $\mathbf{v} \mapsto \mathbf{v}B$ is a rotation through ϕ about the y-axis. $-\tan\theta = a_2/a_1$. $-\tan\phi = \sqrt{(a_1^2 + a_2^2)}/a_3$.

16 $\beta^{-1}\alpha\beta$ is in $SO(3)$ because $SO(3)$ is a group. $\beta^{-1}\alpha\beta$ has eigenvector $(0, 0, 1)$.

17 $-I$.

18 Because $SO(3)$ has index 2 in $O(3)$, every element in $O(3) - SO(3)$ has the form $-A$ for some $A \in SO(3)$.

19 $\alpha\tau^{-1}$ is an isometry fixing $\mathbf{O}$. so A is orthogonal from ans 5 and 7.

20 $\alpha\tau^{-1}$ is an orthogonal transformation from qns 5 and 7. So $\alpha{:}\mathbf{v} \mapsto \mathbf{v}A + \mathbf{c}$ where A is orthogonal.

21 $\alpha_i\alpha_j$ is in the group by closure. $\alpha_1\alpha_j = \alpha_2\alpha_j \Rightarrow \alpha_1 = \alpha_2$. So the elements of the second set are distinct. The orbit of $\mathbf{v}$ consists of the $\mathbf{v}_i$. $\mathbf{v}_1\alpha_j = \mathbf{v}\alpha_1\alpha_j$.

22 $\mathbf{w}\alpha = \dfrac{1}{n}(\mathbf{v}_1A + \mathbf{v}_2A + \ldots + \mathbf{v}_nA) + \mathbf{c}$

$\qquad = \dfrac{1}{n}[\mathbf{v}_1A + \mathbf{c} + \mathbf{v}_2A + \mathbf{c} + \ldots + \mathbf{v}_nA + \mathbf{c}]$

$\qquad = \dfrac{1}{n}[\mathbf{v}_1\alpha + \mathbf{v}_2\alpha + \ldots + \mathbf{v}_n\alpha]$

$\qquad = \mathbf{w}.$

23 If G has the fixed point $\mathbf{c}$ and $\mathbf{O}\tau = \mathbf{c}$ then the group $\tau G\tau^{-1}$ stabilises $\mathbf{O}$.

24 A rotation through $k\theta_2$ is in the group by closure. If $\theta_i \neq k\theta_2$ for any k, then for some k, $k\theta_2 < \theta_i < (k + 1)\theta_2$ and $0 < \theta_i - k\theta_2 < \theta_2$. Then there would be a rotation through an angle $< \theta_2$ in the group. Contradiction. This finite subgroup is generated by the rotation through θ_2.

25 The image of G under $A \mapsto \det A$ is $\{\pm 1\}$. The kernel, $G \cap SO(2)$, has two cosets in G.

26 $\beta\alpha\beta = \alpha^{-1}$. If G is a finite subgroup of $O(2)$ and $G \cap SO(2)$ has order n, by qn 24 $G \cap SO(2)$ is cyclic. Let α be a generator of this group and let β be a reflection in G, then $\langle \alpha, \beta \rangle$ is a dihedral group by qn 6.43 and $\langle \alpha, \beta \rangle$ exhausts G by qn 25.

27 Order of G is 6. Two green points in one orbit. Six red points in two orbits.

28 Six blue points in one orbit. Eight green points in one orbit. Twelve red points in one orbit.

30 If P is an m-pole, the length of the orbit containing P is $|G|/m$. If Q is any other point in the orbit, the order of its stabiliser is $|G|/(|G|/m) = m$ so Q is an m-pole. Alternatively, if α is a rotation of order m with pole P and $P\beta = Q$ then $\beta^{-1}\alpha\beta$ is a rotation of order m with pole Q.

31 $m - 1$.

32 Because each rotation has two poles.

34 $N \geqslant 2 \Rightarrow 0 < \dfrac{2}{N} \leqslant 1 \Rightarrow -1 \leqslant -\dfrac{2}{N} < 0 \Rightarrow 1 \leqslant 2 - \dfrac{2}{N} < 2.$

$\qquad 1 - \dfrac{1}{m} < 1 \leqslant 2 - \dfrac{2}{N}$

so only one orbit is impossible. If there were four or more orbits,

$$\sum\left(1 - \frac{1}{m}\right) \geqslant 4 \cdot \frac{1}{2} = 2 > 2 - \frac{2}{N}.$$

36 If

$$l, m, n \geqslant 3,$$

then

$$\frac{1}{l}, \frac{1}{m}, \frac{1}{n} \leqslant \frac{1}{3},$$

so

$$\frac{1}{l} + \frac{1}{m} + \frac{1}{n} \leqslant 1 < 1 + \frac{2}{N}.$$

37 $\dfrac{1}{2} + \dfrac{1}{2} + \dfrac{1}{n} = 1 + \dfrac{2}{N} \Rightarrow n = \dfrac{1}{2} N.$

38 $\dfrac{1}{2} + \dfrac{1}{3} + \dfrac{1}{n} = 1 + \dfrac{2}{N} \Rightarrow n = \dfrac{6N}{N + 12}.$

Since $n \geqslant m, n \geqslant 3$. Since

$$n = \frac{6N}{N + 12}, \quad n < 6.$$

$n = 3 \Rightarrow N = 12.$ $n = 4 \Rightarrow N = 24.$ $n = 5 \Rightarrow N = 60.$

39 $m, n \geqslant 4 \Rightarrow \dfrac{1}{m} + \dfrac{1}{n} \leqslant \dfrac{1}{2} \Rightarrow \dfrac{1}{2} + \dfrac{1}{m} + \dfrac{1}{n} \leqslant 1 < 1 + \dfrac{2}{N}.$

40 For $A \mapsto \det A$, kernel $= G \cap SO(3)$. Because $\det A = \pm 1$, the kernel has either one or two cosets.

41 $e = (1)(a)$, $\alpha = (1234)(abcd)$, $\alpha^2 = (13)(24)(ac)(bd)$, $\alpha^3 = (1432)(adcb)$, $\beta = (1a)(2d)(3c)(4b)$, $\beta\alpha = (1b)(2a)(3d)(4c)$, $\beta\alpha^2 = (1c)(2b)(3a)(4d)$, $\beta\alpha^3 = (1d)(2c)(3b)(4a)$. Reflections: $\alpha^2\gamma = (1a)(2b)(3c)(4d)$, $\beta\gamma = (13)(ac)$, $\beta\alpha\gamma = (14)(23)(ad)(bc)$, $\beta\alpha^2\gamma = (24)(bd)$, $\beta\alpha^3\gamma = (12)(34)(ab)(cd)$. Not reflections: $\gamma = (1c)(2d)(3a)(4b)$, $\alpha\gamma = (1d3b)(2a4c)$, $\alpha^3\gamma = (1b3d)(2c4a)$. γ has matrix $-I$ and so is in the centre of the group. Use qn 10.11.

42 $S_4 \times C_2.$

43 $e = (1)(a)$, $\alpha = (123)(abc)$, $\alpha^2 = (132)(acb)$, $\beta = (1a)(2c)(3b)$, $\beta\alpha = (1b)(2a)(3c)$, $\beta\alpha^2 = (1c)(2b)(3a)$. Reflections: $\delta^3\gamma = (1a)(2b)(3c)$, $\beta\delta^3\gamma = (23)(bc)$, $\beta\delta\gamma = (13)(ac)$, $\beta\delta^5\gamma = (12)(ab)$. Not reflections: $\delta^5\gamma = (1b3a2c)$, $\delta\gamma = (1c2a3b)$. The full symmetry group of the prism D_6 is generated by $\delta\gamma$ of order 6 and β of order 2. On the twelve points, $\beta = (1a)(2c)(3b)(1'a')(2'c')(3'b')$ so $\beta\delta\beta = \delta^{-1}$ and $\langle \delta, \beta \rangle$ is a dihedral group of rotations.

44 Argue as in qn 18.

45 If $\alpha_i = \beta_j$ for some i, j, then the $\beta_j \gamma$ form the coset $[G \cap SO(3)]\gamma$. If the α_i are distinct from the β_i there are $2n$ in total. $\alpha_i \beta_j = \alpha_i(\beta_j \gamma)\gamma = (\beta_k \gamma)\gamma = \beta_k$. $\beta_i \beta_j = (\beta_i \gamma)(\beta_j \gamma) = \alpha_k$, so the set is closed and forms a group.

22

Discrete groups fixing a line

Discrete groups are the groups of symmetries of ornamental designs. The discrete groups fixing a point are the groups of symmetries of ornaments with a centre. In the plane these must be finite cyclic or dihedral groups. The discrete groups fixing a line are the groups of symmetries of ribbons or friezes.

To study the frieze groups we first enumerate all the isometries fixing a line with the algebra of complex numbers. We then look at the image of this group under a particular homomorphism which maps any isometry to an isometry with a fixed point. The subgroups we can obtain as images are the basis of our classification.

Concurrent reading: Martin, chapter 10; Lockwood and Macmillan, chapters 3, 16; Coxeter (1969), p. 48; Burn (1973).

Isometries fixing a line

1 For each of the seven patterns given here, presuming that each extends to infinity both to the left and to the right, name the types of isometries in the symmetry group of each pattern.

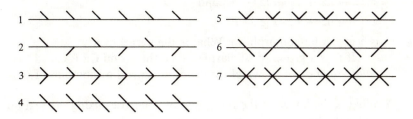

2 If the isometries $z \mapsto e^{i\theta}z + c$ and $z \mapsto e^{i\theta}\bar{z} + c$ each map the real line onto itself, determine a necessary restriction on the value of c by considering the image of 0. By further considering the image of 1, determine a necessary restriction on the value of $e^{i\theta}$.

3 Give geometric descriptions of the transformations

$$z \mapsto z + c,$$
$$z \mapsto -z + c,$$
$$z \mapsto -\bar{z} + c,$$
$$z \mapsto \bar{z} + c,$$

for real values of c. Deduce that the necessary conditions on c and $e^{i\theta}$ in qn 2 for isometries to fix the real line are in fact sufficient conditions. Does the set of all isometries of these types form a group?

4 Prove that the reflection $z \mapsto \bar{z}$ lies in the centre of the group of isometries fixing the real line.

5 Let G be a group of isometries fixing the real line, but not stabilising any point.
   (i) If G contains a half-turn α with centre A and β is an isometry in G which moves A, describe $\beta^{-1}\alpha\beta$ geometrically, and hence describe $\alpha(\beta^{-1}\alpha\beta)$ geometrically.
   (ii) If G contains a reflection with axis perpendicular to the real line, devise an argument similar to that of (i) to prove that G must contain a translation.
   (iii) If G contains a glide-reflection, why must G contain a translation? Deduce that G must contain a translation.

A homomorphic image: the point group

6 Establish that the mapping π of the group of similarities defined by

$$\pi : [z \mapsto az + b] \mapsto [z \mapsto az] \text{ and}$$
$$\pi : [z \mapsto a\bar{z} + b] \mapsto [z \mapsto a\bar{z}]$$

is a group homomorphism. What is the kernel of π?

   If G is a group of similarities, then the group $G\pi$ is called the *point-group* of G. (Beware! $G\pi$ is not in general a subgroup of G.)

7 What is the image of the full group of isometries fixing the real line under the homomorphism π of qn 6?

8 If G is a subgroup of the full group of isometries fixing the real line and $G\pi$ (as in qn 6) consists of the identity alone, what can be said about the elements of G?

9 If G is a subgroup of the group of isometries fixing the real line and $G\pi = \{[z \mapsto z], [z \mapsto -z]\}$, what can be said about the elements of G?

10 If G is a subgroup of the group of isometries fixing the real line and $G\pi = \{[z \mapsto z], [z \mapsto -\bar{z}]\}$, what can be said about the elements of G?

When discussing symmetries of the plane, we distinguish between the isomorphic groups C_2 and D_1 by presuming that C_2 contains a half-turn and D_1 contains a reflection. Thus the point group in qn 9 is C_2, and the point group in qn 10 is D_1.

11 If G is a subgroup of the full group of isometries fixing the real line and $G\pi = \{[z \mapsto z], [z \mapsto \bar{z}]\}$, give an example to show that although the point group is D_1, the group G need not contain a reflection.

12 If G is a subgroup of the full group of isometries fixing the real line and $G\pi = D_2$, give an example to show that G need not contain a reflection in the real line. First choose a suitable pattern and then describe its symmetries.

Discrete groups of transformations

13 A group of transformations of the plane to itself is said to be *discrete* when, for each point P of the plane, it is possible to draw a circle centre P within which there are no other points of the orbit of P. Which of the following groups are discrete:
(a) the full group of rotations with centre 0,
(b) the full group of translations fixing the real line,
(c) D_6, being the full symmetry group of a regular hexagon in the plane?

14 If a discrete group of isometries is given which stabilises a point O, by considering the orbit of a point other than O, establish that if the group contains any rotations distinct from the identity then it contains a rotation through a minimum angle. Use qns 21.24 and 21.26 to show that the group is finite, and either cyclic or dihedral.

15 The *length* of the translation $z \mapsto z + c$ is defined to be $|c|$. If a discrete group of isometries of the plane is given which contains a translation, prove that the group contains a translation of minimal length.

16 Deduce from qns 5 and 15 that a discrete group which fixes a line and does not stabilize a point contains a translation of minimal length.

Classification of frieze groups

17 If a discrete group of isometries fixes a line but does not stabilise a point, it is called a *frieze group*. A pattern whose group of symmetries is a frieze group is called a frieze pattern, or simply a frieze.
Exhibit a frieze pattern for which the frieze group contains only translations. Must this frieze group be cyclic?
The symbol C_∞ is sometimes used to denote an infinite cyclic group.

18 Exhibit a frieze pattern for which the frieze group contains half-turns, but neither reflections nor glide-reflections.
If a minimal translation in this group is $\tau:z \mapsto z + 1$ and α is a half-turn in the group, verify that $\alpha\tau\alpha = \tau^{-1}$. Because α^2 is the identity, this relation justifies referring to $\langle \tau, \alpha \rangle$ as D_∞. By choosing the origin at the centre of α, exhibit all the elements of this frieze group as mappings of complex numbers.

19 Exhibit a frieze pattern for which the frieze group contains reflections with axes perpendicular to the fixed line, but neither half-turns nor glide-reflections.
If a minimal translation in this group is $\tau:z \mapsto z + 1$ and ϱ is a reflection in the group, verify that $\varrho\tau\varrho = \tau^{-1}$. Because ϱ^2 is the identity, this relation justifies referring to $\langle \tau, \varrho \rangle$ as D_∞. By choosing the imaginary axis as the axis of ϱ, exhibit all the elements of this frieze group as mappings of complex numbers. If the imaginary axis is not the axis of ϱ and $\varrho:z \mapsto -\bar{z} + r$ for some real number r, determine all the elements of $\langle \tau, \varrho \rangle$ in this case as mappings of complex numbers.

20 If a frieze group contains a minimal translation $z \mapsto z + 1$, what glide-reflections may be in the group?
If a frieze group contains only translations and glide-reflections, prove that it is a cyclic group, and exhibit a frieze with this group of symmetries.

21 If a frieze group contains any two of the three types of isometry,
a half-turn,
a glide-reflection,
a reflection with axis perpendicular to the fixed line,
prove that it contains all three types of isometry.
Exhibit a frieze pattern for which the frieze group includes

isometries of these three types, but not a reflection in the fixed line. If $z \mapsto z + 1$ is a minimal translation in this group and $z \mapsto -z$ is a half-turn in the group, exhibit all the elements of the group as mappings of complex numbers. By a suitable choice of generators, show that this group is abstractly isomorphic to D_∞.

22 If a frieze group contains a minimal translation $z \mapsto z + 1$, the reflection $z \mapsto \bar{z}$ and no other reflections, prove that it may not contain any half-turns, and exhibit all the elements of the group as mappings of complex numbers. Show that this group is abstractly isomorphic to $C_2 \times C_\infty$. Use qns 4 and 10.11. Draw a frieze pattern with such a frieze group.

23 If a frieze group contains a minimal translation $\tau : z \mapsto z + 1$ and the reflection $\varrho : z \mapsto \bar{z}$, and contains some other isometry not in $\langle \varrho, \tau \rangle$, prove that it contains a half-turn, and by taking the origin at the centre of this half-turn, prove that every transformation of the form $z \mapsto \pm z + n$ or $z \mapsto \pm \bar{z} + n$ lies in the group, when n is an integer, and that these exhaust the elements of the group.
Use qns 4 and 10.11 to show that this group is abstractly isomorphic to $C_2 \times D_\infty$. Draw a frieze pattern with such a frieze group.

24 For each of the following frieze patterns determine an abstract group to which the frieze group is isomorphic and also the related point group (as in qn 6).

   (i) $\cdots$LLLLLLL$\cdots$ (ii) $\cdots$ZZZZZZZ$\cdots$ (iii) $\cdots$VVVVVVV$\cdots$
  (iv) $\ldots$D$_D$D$_D$D$_D$D$\ldots$ (v) $\ldots$I$_I$I$_I$I$\ldots$ (vi) $\cdots$DDDDDDD$\cdots$
 (vii) $\cdots$IIIIIII$\cdots$

Summary

Theorem qn 6	The mapping of the group of similarities given by $[z \mapsto az + b] \mapsto [z \mapsto az]$ and $[z \mapsto a\bar{z} + b] \mapsto [z \mapsto a\bar{z}]$ is a homomorphism. The image of a group G under this homomorphism is called the *point-group* of G.
Definition qn 13	A group of transformations of the plane to itself is said to be *discrete* when for any point P of the plane, a circle may be drawn with centre P which contains no other point of the orbit of P.
Theorem qn 14	A discrete group which stabilises a point of the plane is a finite group, and either cyclic or dihedral.
Definition qn 17	A discrete group which fixes a line but does not stabilise a point is called a *frieze group*.

Theorem qns 16, 17	A frieze group contains a cyclic translation subgroup.
Theorem qn 7	The point group of a frieze group is either C_1, C_2, D_1 or D_2.
Theorem qn 17	The frieze group with point group C_1 is a cyclic group consisting wholly of translations.
Theorem qn 18	The frieze group with point group C_2 is an infinite dihedral group consisting of half-turns and translations.
Theorem qns 19, 20, 22	There are three frieze groups with point group D_1. One is a cyclic group generated by a glide-reflection. One is a dihedral group consisting of reflections and translations. The third is a direct product of subgroups generated by a translation and a reflection in the axis.
Theorem qns 21, 23	There are two frieze groups with point group D_2. One is an infinite dihedral group generated by a glide-reflection and a half-turn. The other contains all the possible transformations and is the direct product of the infinite dihedral group generated by a translation and a half-turn and the subgroup generated by a reflection in the axis.

Historical note

The first intensive use of complex numbers to analyse groups of isometries in the plane appears in R. Fricke and F. Klein *Vorlesungen über die Theorie der automorphen Functionen*, vol. I (1897). In this text, discrete groups are called 'groups without infinitesimal substitutions'. The papers of G. Polya and P. Niggli (1924) reminded mathematicians of the discrete groups in the plane, and the first account of the frieze groups appears in the second edition of A. Speiser *Theorie der Gruppen von endliche Ordnung* (1927).

Answers to chapter 22

1 All have translations. Glide-reflections 2, 3, 6, 7. Reflections 3, 5, 6, 7. Half-turns 4, 6, 7.

2 c is real and $e^{i\theta}$ is real so $e^{i\theta} = \pm 1$.

3 $z \mapsto z + c$, translation. $z \mapsto -z + c$, half-turn. $z \mapsto -\bar{z} + c$, reflection. $z \mapsto \bar{z} + c$, reflection if $c = 0$, glide-reflection otherwise.

4 The crucial fact is that $\bar{c} = c$ when c is real.

5 (i) $\beta^{-1}\alpha\beta$ is a half-turn with centre $A\beta$. So $\alpha(\beta^{-1}\alpha\beta)$ is a translation through twice the distance $AA\beta$. (ii) If α is a reflection with axis l perpendicular to the real axis and β moves l, then $\beta^{-1}\alpha\beta$ is a reflection with axis $l\beta$. (iii) If γ is a glide-reflection γ^2 is a translation.

6 The translation group is the kernel of π.

7 $\{[z \mapsto \pm z], [z \mapsto \pm \bar{z}]\}$.

8 All are translations.

9 All are translations or half-turns.

10 All are translations or reflections with axes perpendicular to the real line.

11 Group generated by the glide-reflection $z \mapsto \bar{z} + 1$.

12 Pattern 6 of qn 1. The group generated by $z \mapsto -z$ and $z \mapsto \bar{z} + 1$ contains $z \mapsto \pm z + 2n$ and $z \mapsto \pm \bar{z} + 2n + 1$ for all $n \in \mathbb{Z}$.

13 Only (c).

14 If the rotations about O can be through arbitrarily small angles, then the orbits cluster.

15 If the translations can be of arbitrarily small length, then the orbits cluster.

16 If no point is stabilized there must be either a translation or a glide-reflection in the group.

17 Pattern 1 of qn 1. The group is generated by $z \mapsto z + 1$, a translation of minimal length.

18 Pattern 4 of qn 1. $z \mapsto \pm z + n$.

19 Pattern 5 of qn 1. $z \mapsto z + n$, $z \mapsto -\bar{z} + n$. $z \mapsto z + n$, $z \mapsto -\bar{z} + r + n$.

20 $z \mapsto \bar{z} + n$ and $z \mapsto \bar{z} + n + \frac{1}{2}$. If no reflections then $z \mapsto \bar{z} + n$ is not admissable, so group consists of $z \mapsto z + n$ and $z \mapsto \bar{z} + n + \frac{1}{2}$ for all $n \in \mathbb{Z}$. Pattern 2 of qn 1.

21 Half-turn $z \mapsto -z + c$. Glide-reflection $z \mapsto \bar{z} + d$ and reflection

$z \mapsto -\bar{z} + e$. The numbers c, d, e are real. Product of any two gives third, except possibly reflection and half-turn may give $z \mapsto \bar{z}$. Then with translations, glide-reflections exist. Pattern 6 of qn 1. $z \mapsto \pm z + n$, $z \mapsto \pm \bar{z} + n + \frac{1}{2}$. If $\alpha : z \mapsto \bar{z} + \frac{1}{2}$ and $\beta : z \mapsto -z$ then $\beta^{-1}\alpha\beta = \alpha^{-1}$ and the group is dihedral.

22 Product of half-turn and $z \mapsto \bar{z}$ is a reflection with axis perpendicular to the real axis. So all elements have the form $z \mapsto z + n$ or $z \mapsto \bar{z} + n$. The reflection $z \mapsto \bar{z}$ is in the centre and the translations are generated by $z \mapsto z + 1$. Pattern 3 of qn 1.

23 Isometries not in $\langle \varrho, \tau \rangle$ are either half-turns, reflections perpendicular to the real axis or glide-reflections. If α is a reflection with axis perpendicular to the real axis, $\alpha\varrho$ is a half-turn. If γ is a glide-reflection not in $\langle \varrho, \tau \rangle$ then $\gamma : z \mapsto \bar{z} + r$ with r not an integer, but then the group contains the translation $\varrho\gamma\tau^{-[r]} : z \mapsto z + r - [r]$, contradicting the minimality of $z \mapsto z + 1$. The group of qn 22 with $z \mapsto -z$ is of the required form. Pattern 7 of qn 1.

24

	(i)	(ii)	(iii)	(iv)	(v)	(vi)	(vii)
Point group	C_1	C_2	D_1	D_1	D_2	D_1	D_2
Full group	C_∞	D_∞	D_∞	C_∞	D_∞	$C_\infty \times C_2$	$D_\infty \times C_2$

23

Wallpaper groups

The key step towards classifying discrete groups in the plane not fixing a point or a line is to establish the crystallographic restriction, that is, that the only possible rotations have order 2, 3, 4, or 6. The consequence of this is that a wallpaper group has one of ten different point groups, C_1, C_2, C_3, C_4, C_6, D_1, D_2, D_3, D_4 or D_6. The analysis of each of these ten possibilities, which is in some cases laborious, completes the classification.

Concurrent reading: *School Mathematics Project*; Lockwood and Macmillan, chapters 12, 18; Schattschneider; Martin, chapter 11.

Crystallographic restriction

1 A group G of isometries of the plane contains a translation τ and a rotation α with centre A through an angle θ. Find the image of $A\tau$ under $\tau^{-1}\alpha^{-1}\tau\alpha$ and show that the length of the translation $\tau^{-1}\alpha^{-1}\tau\alpha$ is $2\sin\frac{1}{2}\theta$ times the length of the translation τ (first use qn 17.9 to find $\alpha^{-1}\tau\alpha$).
   If G is a discrete group and τ is a minimal translation in G (qn 22.16) deduce that $2\sin\frac{1}{2}\theta \geqslant 1$ and so $\theta \geqslant \frac{1}{3}\pi$.

2 A group G of isometries of the plane contains a translation τ and a rotation α with centre A through an angle $\frac{2}{5}\pi$. Find the image of $A\tau^{-1}$ under $\tau\alpha^{-2}\tau\alpha^2$ and show that the length of the translation $\tau\alpha^{-2}\tau\alpha^2$ is $2\sin\frac{1}{10}\pi$ times the length of the translation τ.
   Deduce that G cannot be a discrete group.

3 If a discrete group of isometries of the plane contains a translation and a rotation, prove that the order of that rotation is 2, 3, 4 or 6.
   This is known as the *crystallographic restriction*.

Possible point groups

4 If a discrete group of isometries of the plane contains a translation, prove that its point group (qn 22.6) is either C_1, C_2, C_3, C_4, C_6, D_1, D_2, D_3, D_4 or D_6.

5 When a discrete group of isometries of the plane does not stabilise either a point or a line is called a *wallpaper group*. If a wallpaper group W contains a rotation α and β is an element of W which moves the centre of α, prove that either $\alpha^{-1}\beta^{-1}\alpha\beta$ (if β is direct) or $\alpha\beta^{-1}\alpha\beta$ (if β is opposite) is a translation in W by considering the images of these elements in the point group.

6 If a wallpaper group W contains a reflection ϱ and β is an element of W which moves the axis of ϱ, prove that $\varrho\beta^{-1}\varrho\beta$ is either a translation or a rotation. Prove that every wallpaper group contains a translation.

Point group C_1

7 Let W be a wallpaper group with point group C_1, then W consists wholly of translations. Let τ be a translation of minimal length in W and let σ be a translation of minimal length in $W - \langle\tau\rangle$. Describe the orbit of any point O under the subgroup $\langle\tau, \sigma\rangle$ of W. If W contains a translation not in $\langle\tau, \sigma\rangle$ show that the orbit of O under W contains a point inside the triangle O, $O\tau$, $O\sigma$. Must such a point contradict the minimality of τ or σ? Deduce that W is isomorphic to the direct product of two cyclic groups.
Such a wallpaper group is denoted by *p1*.

Point group C_2

8 Let W be a wallpaper group with point group C_2, then W consists wholly of half-turns and translations. If α is a half-turn in W and T is the translation group of W, why must $W = T \cup T\alpha$? What condition for wallpaper groups would fail if T only contained

translations in one direction? Deduce that T is a wallpaper group, and that if $T = \langle \tau, \sigma \rangle$, then $W = \langle \alpha, \tau, \sigma \rangle$.
Such a wallpaper group is denoted by *p2*.

9 If a wallpaper group contains a rotation of order n, then any point which is the centre of a rotation of order n is known as an *n-centre* for the group. If a wallpaper group is of the type *p2*, how many distinct orbits of 2-centres does it have?

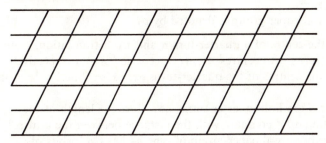

Point group C_3

10 Let W be a wallpaper group with point group C_3, then W consists wholly of rotations through $\frac{2}{3}\pi$ or $\frac{4}{3}\pi$ and translations. If α and β are rotations of order 3 in W and T is the translation group of W, must either $\beta\alpha^{-1}$ or $\beta\alpha^{-2} \in T$? Deduce that $W = \langle \alpha, T \rangle$. If τ is a translation of minimal length in W, use qn 7 to show that $T = \langle \tau, \alpha^{-1}\tau\alpha \rangle$, and deduce that $W = \langle \alpha, \tau \rangle$.
Such a wallpaper group is denoted by *p3*.

11 If A is the centre of a rotation α through $\frac{2}{3}\pi$ and τ is a translation, name rotations with centres $A\tau$, $A\tau\alpha$, $A\tau\alpha^2$, $A\tau^{-1}$, $A\tau^{-1}\alpha$ and $A\tau^{-1}\alpha^2$ respectively. Locate the centres of the rotations $\alpha\tau$, $\tau\alpha$, $\alpha\tau^{-1}$, $\tau^{-1}\alpha$, $\alpha^{-1}\tau\alpha^{-1}$ and $\alpha^{-1}\tau^{-1}\alpha^{-1}$ in this hexagon. If α and β are both rotations through an angle $\frac{2}{3}\pi$, how does the length of the translation $\alpha\beta^{-1}$ compare with the distance between the centres?
Deduce that a wallpaper group of the type *p3* has three orbits of 3-centres.

Point group C_4

12 Let W be a wallpaper group with point group C_4, then W consists wholly of rotations through $\frac{1}{2}\pi$, π, $\frac{3}{2}\pi$ and translations. If α is a rotation of order 4 in W, β is any rotation in W and T is the translation group of W, must either $\beta\alpha^{-1}$, $\beta\alpha^{-2}$ or $\beta\alpha^{-3} \in T$? Deduce that $W = \langle \alpha, T \rangle$. If τ is a translation of minimal length in W, use qn 7 to show that $T = \langle \tau, \alpha^{-1}\tau\alpha \rangle$, and deduce that $W = \langle \alpha, \tau \rangle$.
Such a wallpaper group is denoted by *p4*.

13 If A is the centre of a quarter-turn α and τ is a translation, name quarter-turns with centres $A\tau$, $A\tau^{-1}$, $A\tau\alpha$ and $A\tau^{-1}\alpha$ respectively. Locate the centres of the quarter-turns $\alpha\tau$, $\tau\alpha$, $\alpha\tau^{-1}$ and $\tau^{-1}\alpha$ in this square.
If α and β are both quarter-turns, how does the length of the translation $\alpha\beta^{-1}$ compare with the distance between the centres? Deduce that a wallpaper group of type *p4* has two orbits of 4-centres.

14 Prove that a wallpaper group of type *p4* has exactly one orbit of 2-centres which are not 4-centres.

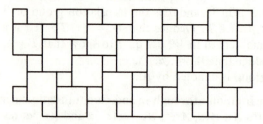

Point group C_6

15 Let W be a wallpaper group with point group C_6, then W consists wholly of rotations through $\frac{1}{3}\pi$, $\frac{2}{3}\pi$, π, $\frac{4}{3}\pi$, $\frac{5}{3}\pi$ and translations. If α is a rotation of order 6 in W, β is any rotation in W and T is the translation group of W, must either $\beta\alpha^{-1}$, $\beta\alpha^{-2}$, $\beta\alpha^{-3}$, $\beta\alpha^{-4}$ or $\beta\alpha^{-5} \in T$? Deduce that $W = \langle \alpha, T \rangle$. If τ is a translation of minimal length in W, use qn 7 to show that $T = \langle \tau, \alpha^{-1}\tau, \alpha \rangle$, and deduce that $W = \langle \alpha, \tau \rangle$.
Such a wallpaper group is denoted by *p6*.

16 If α and β are both rotations through $\frac{1}{3}\pi$ with centres A and B respectively, verify that $\alpha\beta^{-1}\alpha^{-1}\beta$ is a translation mapping A to B. Deduce that in a wallpaper group of type *p6* that the 6-centres lie in a single orbit under the translation group. By considering the transformation $(\alpha\beta^{-1}\alpha^{-1}\beta)\beta^3$ prove that the midpoint of AB is a

2-centre. Prove that every 2-centre in such a group lies at the mid-point of the line joining two 6-centres by considering the half-turn η and the rotations α and $\eta\alpha\eta$.

17 If α and β are both rotations through $\frac{1}{3}\pi$ with centres A and B respectively, determine the nature of the transformation $\beta\alpha$. Deduce that the 3-centres which are not 6-centres lie in a single orbit of a wallpaper group of type *p6*.

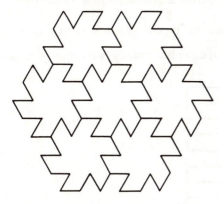

Point group D_1

18 What is the axis of $z \mapsto e^{i\theta}\bar{z}$? Show that all the reflections and glide-reflections which are mapped to the same element of the point group have parallel axes.

19. Let W be a wallpaper group with point group D_1, then W consists of opposite isometries and translations. If γ is an opposite isometry in W and T is the translation group of W, why must $W = T \cup T\gamma$? Let l be the axis of γ. What condition for wallpaper groups would fail if T only contained translations in the direction of l? Let α be a translation in W such that $l \neq l\alpha$. Let $\gamma = \beta\varrho$, where ϱ is the reflection in l and β is either the identity or a translation in the direction of l. (Note, we do not assume that either β or ϱ necessarily belongs to W.) Prove that $\alpha^{-1}\gamma^{-1}\alpha\gamma = (\alpha^{-1}\varrho\alpha)\varrho$ and deduce that W contains translations perpendicular to l.

If γ is a glide-reflection, why must W contain translations in the direction of l? If γ is the reflection in l and T only contained translations perpendicular to l what condition for wallpaper groups would fail? If γ is the reflection in l and μ is a translation in W moving a line perpendicular to l, prove that $(\gamma\mu)^2$ is a translation. Further, by considering the image of a point on l under the two translations μ and $\gamma\mu\gamma$, deduce that the image of this point under $\gamma\mu\gamma\mu$ lies on l, so that W contains translations in the direction of l.

20 Let W be a wallpaper group with point group D_1 containing an opposite isometry γ with axis l. Let τ be a minimal translation in W in the direction of l and let σ be a minimal translation in W perpendicular to l. Such translations exist by qn 19. Now we suppose that $\langle \tau, \sigma \rangle$ is the translation group of W. Why must $\gamma^2 \in \langle \tau \rangle$? If $\gamma^2 = \tau^m$ and m is even, construct a reflection ϱ in W. In this case $W = \langle \varrho, \tau, \sigma \rangle$ and this type of group is known as *pm*. If $\gamma^2 = \tau^m$ and m is odd, construct a glide-reflection δ such that $\langle \delta \rangle = \langle \gamma, \tau \rangle$.
In this case $W = \langle \delta, \sigma \rangle$ and this type of group is known as *pg*.

21 Prove that a group of the type *pm* has exactly two orbits of axes of reflections.

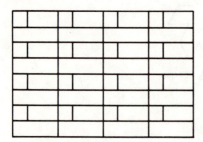

22 Prove that a group of the type *pg* contains no reflections.

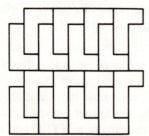

23 Let W be a wallpaper group with point group D_1 containing an opposite isometry γ with axis l. Let τ be a minimal translation in W in the direction of l and let σ be a minimal translation in W perpendicular to l. We suppose that $\langle \tau, \sigma \rangle$ is a *proper* subgroup of the translation group of W.

Use the argument of qn 7 to show that the translation group of W contains an element μ such that $O\mu$ lies in the triangle O, $O\tau$, $O\sigma$, but not on either of the sides $OO\tau$ or $OO\sigma$. If O is a point on l show that the four points $O\mu$, $O\mu^{-1}$, $O\gamma^{-1}\mu\gamma$ and $O\gamma^{-1}\mu^{-1}\gamma$ lie at the vertices of a rectangle for which l is a line of symmetry. Deduce that the translation $\gamma^{-1}\mu^{-1}\gamma\mu \in \langle \sigma \rangle$ and the translation $\gamma^{-1}\mu\gamma\mu \in \langle \tau \rangle$ and from the position of $O\mu$ in the triangle O, $O\tau$,

$O\sigma$ that in fact $\gamma^{-1}\mu^{-1}\gamma\mu = \sigma$ and $\gamma^{-1}\mu\gamma\mu = \tau$. From the first of these equations deduce that $O\mu$ lies on the median of O, $O\tau$, $O\sigma$ parallel to $OO\tau$, and from the second of these equations that $O\mu$ lies on the median parallel to $OO\sigma$. Hence $O\mu$ is the midpoint of $O\tau O\sigma$ and $\mu^2 = \tau\sigma$. Use the uniqueness of μ to show that the translation group of W is $\langle \mu, \tau, \sigma \rangle = \langle \mu, \tau \rangle = \langle \mu, \sigma \rangle$.

As in qn 20, $\gamma^2 \in \langle \tau \rangle$. If $\gamma^2 = \tau^m$ and m is even, construct a reflection with axis l in W. If $\gamma^2 = \tau^m$ and m is odd, and we let $\delta = \gamma\tau^{\frac{1}{2}(1-m)}$ be the glide-reflection such that $\langle \delta \rangle = \langle \gamma, \tau \rangle$, find a point which is fixed by $\mu^{-1}\delta$, so that this opposite isometry in W must be a reflection. Whether m is even or odd, W contains a reflection ϱ and $W = \langle \varrho, \mu, \tau \rangle$.
This type of group is known as *cm*.

24 Under a group of type *cm* prove that the axes of reflections all lie in one orbit and that the axes of glide-reflections which are not axes of reflections also lie in one orbit.

Point group D_2

25 Let W be a wallpaper group with point group D_2, then W consists of translations, half-turns and opposite isometries. Let γ and δ be opposite isometries in W with perpendicular axes. By considering the image of $\gamma\delta$ in the point group, prove that $\gamma\delta$ is a half-turn and deduce that $W = \langle \gamma, \delta, T \rangle$ where T is the translation group of W.

Use qn 19 to show that there exists a minimal translation τ in the direction of the axis of γ and a minimal translation σ in the direction of the axis of δ.

Suppose now that $T = \langle \tau, \sigma \rangle$ and that $\gamma^2 = \tau^m$ and $\delta^2 = \sigma^n$.
(i) If both m and n are even, find reflections ϱ_1 and ϱ_2 in the axes of γ and δ.
In this case $W = \langle \varrho_1, \varrho_2, \tau, \sigma \rangle$ and the group is said to be of the type *pmm*.
(ii) If both m and n are odd, find glide-reflections λ_1 and λ_2 such that $\langle \lambda_1 \rangle = \langle \gamma, \tau \rangle$ and $\langle \lambda_2 \rangle = \langle \delta, \sigma \rangle$.
In this case $W = \langle \lambda_1, \lambda_2 \rangle$ and the group is said to be of the type *pgg*.

(iii) If one of *m* and *n* is even and the other is odd, say *m* is even and *n* is odd, find a reflection ϱ in the axis of γ and a glide-reflection λ such that $\langle\lambda\rangle = \langle\delta, \sigma\rangle$.

In this case $W = \langle\varrho, \lambda, \tau\rangle$ and the group is said to be of the type *pmg*.

26 For a group of the type *pmm* show that there are four orbits of 2-centres, and that every 2-centre lies on the axis of a reflection.

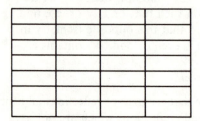

27 For a group of the type *pgg*, show that no 2-centre lies on the axis of a glide-reflection and that there are exactly two orbits of 2-centres.

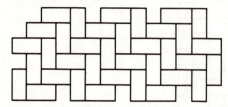

28 For a group of the type *pmg*, show that all 2-centres lie on the axes of glide reflections, that no 2-centres lie on the axes of reflections and that there are exactly two orbits of 2-centres.

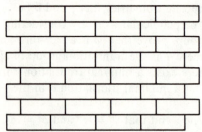

29 Let W be a wallpaper group with point group D_2, and let γ and δ be opposite isometries in W with perpendicular axes. Let τ be a minimal translation in the direction of the axis of γ and let σ be a minimal translation in the direction of the axis of δ. Suppose now that $\langle\tau, \sigma\rangle$ is a *proper* subgroup of the translation group T of W. Use qn 23 to prove that there exists a translation μ in W such that $\mu^2 = \tau\sigma$, and that $T = \langle\mu, \tau\rangle$. Use qn 23 again to prove that W

contains a reflection ϱ_1 with axis parallel to that of γ and a reflection ϱ_2 with axis parallel to that of δ.

In this case $W = \langle \varrho_1, \varrho_2, \mu, \tau \rangle$ and the group is said to be of type *cmm*.

30 For a group of the type *cmm*, show that between two parallel axes of reflections there is an axis of a glide-reflection. Show that there are three orbits of 2-centres, two lying on the axes of reflections and one lying on the axes of glide-reflections.

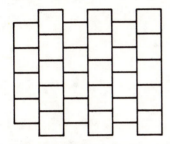

Point group D_3

31 If α is a rotation of order 3 and γ is any opposite isometry whatever, prove that $\gamma\alpha^{-1}\gamma^2\alpha$ is a reflection, by considering a point on the axis of γ.

32 Let W be a wallpaper group with point group D_3, then W consists of rotations of order 3, translations and opposite isometries. Use qn 31 to show that W contains a reflection ϱ, say, and by conjugation by a rotation, show that W contains reflections with axes in all three possible directions. Let A be the point of intersection of two reflection axes of W. Why must A be a 3-centre for W? Let α be the rotation with centre A through an angle $\frac{2}{3}\pi$. If τ is a minimal translation in W use qn 10 to show that $W = \langle \varrho, \alpha, \tau \rangle$.

By qn 19, there exist translations in W parallel to the axis of ϱ. If one of these is minimal in W then the group is said to be of the type *p31m*.

Let σ denote a minimal translation in the direction of the axis of ϱ. Use the argument of qn 7 to prove that if σ is not a translation of minimal length in W, then there exists a translation τ of minimal length such that $A\tau$ lies within the triangle $A, A\sigma, A\sigma^{-1}\alpha^2$. Now use the arguments of qn 23 to show that $\tau(\varrho\tau\varrho) = \sigma$, so that $A\tau$ lies on the perpendicular bisector of $AA\sigma$. Deduce that $A\tau$ is at the centroid of the triangle $A, A\sigma, A\sigma^{-1}\alpha^2$.

In this case $W = \langle \varrho, \alpha, \tau \rangle$ and the group is said to be of type *p3m1*.

33 For a group of type *p31m* show that there are exactly two orbits of 3-centres. Our notation here is that of the International Union of Crystallography and is at odds with that used by Coxeter, and *S.M.P.*

34 For a group of type *p3m1* show that there are three orbits of 3-centres.

Point group D_4

35 Let W be a wallpaper group with point group D_4. Use qn 12 to show that the subgroup of W of direct isometries is $\langle \alpha, \tau \rangle$ where α is a quarter-turn and τ is a translation of minimal length in W. If γ is any opposite isometry in W, why is $W = \langle \gamma, \alpha, \tau \rangle$?

By qn 13, the 4-centres of W lie in two orbits under $\langle \alpha, \tau \rangle$ which we refer to as orbit A (containing A, the centre of α) and orbit B (containing the centre of $\alpha\tau$). What kind of isometry is $\gamma^{-1}\alpha\gamma$? Deduce that $A\gamma$ lies in orbit A or orbit B. If $A\gamma$ lies in orbit A, prove that $A\delta$ lies in orbit A for every opposite isometry δ of W. If $A\gamma$ lies in orbit B, prove that $A\delta$ lies in orbit B for every opposite isometry δ of W.

(i) Suppose now that $A\gamma$ lies in orbit A. If β is a direct isometry in W such that $A\beta = A\gamma$, determine the nature of the isometry $\gamma\beta^{-1}$. If ϱ is a reflection in W with axis through A, locate the possible images of the 4-centre $A\tau$ under the reflections ϱ, $\varrho\alpha$, $\varrho\alpha^2$ and $\varrho\alpha^3$. Deduce that the axes of these four reflections are along, perpendicular to, or inclined at $\frac{1}{4}\pi$ to the direction of τ. When $W = \langle \varrho, \alpha, \tau \rangle$ with the axis of the reflection ϱ through the centre of α and in the direction of τ, W is said to be of the type *p4m*.

(ii) Suppose that $A\gamma$ lies in orbit B, and let B be the centre of the quarter-turn $\alpha\tau = \beta$. Why must there be an opposite isometry in W mapping A to B? Use the four isometries fixing A to construct four distinct opposite isometries which map A to B. Why must the images of B under these four opposite isometries be A, $A\beta$, $A\beta^2$ and $A\beta^3$? Let γ be the opposite isometry in W which maps $A \mapsto B$ and $B \mapsto A\beta^2$. Must γ be a glide-reflection? What is its axis? Check that $(\alpha\gamma)^2 = \tau$, so that $\langle \gamma, \alpha, \tau \rangle = \langle \gamma, \alpha \rangle$. When $W = \langle \gamma, \alpha \rangle$ with the axis of the glide-reflection γ passing through the centre of the quarter-turn α, W is said to be of the type *p4g*.

In a group of type *p4g*, prove that $\alpha^2\gamma$ is a reflection, so that there exist reflections with axes in at least two directions. Why may reflections with axes in four directions not coexist in such a group?

36 In a group of type *p4m* show that there are three orbits of reflection axes and one additional orbit of axes of glide-reflections.

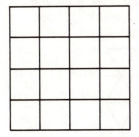

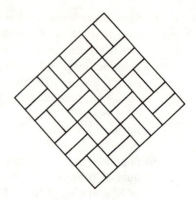

37 In a group of type *p4g* show
that there is exactly one
orbit of reflection axes, one
orbit of glide-reflection axes
containing 4-centres, and
one additional orbit of glide-
reflection axes.

Point group D_6

38 If W is a wallpaper group with point group D_6, use qn 31 to show that
W contains a reflection ϱ. If α is a rotation through $\frac{1}{3}\pi$ in W with
centre A, prove that $A\varrho$ and also $A\varrho\alpha$ are 6-centres for W and so
the axis of ϱ contains a 6-centre. If we suppose now that A lies on
the axis of ϱ, by considering the transformations $\varrho\alpha^n$ for $n = 1, 2,$
3, 4, 5, 6 prove that each of the six possible directions contains
axes of reflections. Use 19 to show that there exist translations in
W parallel to the axis of ϱ. If a minimal such translation is σ and σ
is minimal in W, use qn 15 to show that $W = \langle \varrho, \alpha, \sigma \rangle$. If σ is not
minimal in W use qn 32 to show that there is a minimal translation
τ in W such that $A\tau$ lies at the centroid of the triangle A, $A\sigma$, $A\sigma\alpha$.
Name a reflection with axis $AA\tau$ and deduce that in both cases
$W = \langle \varrho, \alpha, \tau \rangle$ where ϱ is a reflection with axis through the centre
of α in the direction of the minimal translation τ.
Such a group is said to be of type *p6m*.

39 In a group of the type *p6m* show that there is one orbit of 6-centres,
one orbit of 3-centres, one orbit of 2-centres which are not
6-centres and two orbits of axes of reflections.

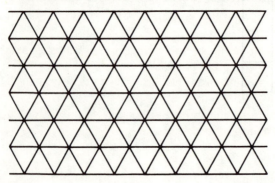

Each of the 17 types is unique up to isomorphism

40 In the case of groups of types *p1, p2, pm, pg, cm, pmm, pmg, pgg* and *cmm*, show that for any two groups of the same type that generators may be made to correspond under conjugation by an affine transformation so that the two translation subgroups, the two subgroups of direct isometries and the two full groups are isomorphic.

41 In the case of groups of the types *p3, p4, p6, p31m, p3m1, p4m, p4g* and *p6m*, show that for any two groups of the same type that generators may be made to correspond under conjugation by a similarity transformation so that the two translation subgroups, the two subgroups of direct isometries and the two full groups are isomorphic.

Summary

Theorem   *The crystallographic restriction.* If a discrete group con-
  qn 3   tains a translation and a rotation, then the rotation is of
         order 2, 3, 4 or 6.

Definition A discrete group which does not stabilise either a point
  qn 5   or a line and which consists of isometries of the plane is
         called a *wallpaper group*.

Theorem   A wallpaper group has point group either C_1, C_2, C_3,
  qn 4 C_4, C_6, D_1, D_2, D_3, D_4 or D_6. The translation subgroup
         of a wallpaper group has two generators.

Theorem Up to isomorphism there is exactly one wallpaper group
qns 7, 8, with point group C_1, C_2, C_3, C_4 or C_6.
10, 12, 15,
40, 41

Theorem Up to isomorphism there are exactly three wallpaper
  qns 20, groups with point group D_1.
  23, 40

Theorem Up to isomorphism there are exactly four wallpaper
  qns 25, groups with point group D_2.
  29, 40

Theorem Up to isomorphism there are exactly two wallpaper
qns 32, 41 groups with point group D_3.

Theorem Up to isomorphism there are exactly two wallpaper
qns 35, 41 groups with point group D_4.

Theorem Up to isomorphism there is exactly one wallpaper group
qns 38, 41 with point group D_6.

Historical note

In 1868, C. Jordan analysed groups of direct isometries of

3-dimensional space, using the word *group* in a geometrical context for the first time. He identified the different kinds of discrete groups of translations, and also, with the help of Bravais's work (1848), the possible point groups. His resulting classification of space groups was incomplete, but it paved the way for the work of E. S. Fedorov (1890) and A. Schönflies (1891). In their study of possible crystal patterns these men showed that there were 230 different discrete groups of 3-dimensional space which do not fix either a point, a line or a plane. Their work stimulated R. Fricke and F. Klein in their first book on automorphic functions (1897) to classify the wallpaper groups. G. Polya acknowledges the stimulus of A. Schönflies' text book in a beautiful little paper which was published in 1924 'Über die Analogie der Kristallsymmetrie in der Ebene'. This paper is a summary of the work of this chapter. Polya classifies the wallpaper groups under their point groups and gives examples of the 17 types of pattern.

In 1900 D. Hilbert asked whether the discrete groups of $\mathbb{R}^n$ were always finite in number. An affirmative answer was provided by L. Bieberbach (1910).

Answers to chapter 23

1 Since the translation group is a normal subgroup, $\alpha^{-1}\tau\alpha$ is a translation, so $\tau^{-1}(\alpha^{-1}\tau\alpha)$ is also a translation.

2 Since the translation group is a normal subgroup, $\alpha^{-2}\tau\alpha^2$ is a translation, so $\tau(\alpha^{-2}\tau\alpha^2)$ is also a translation. The existence of a translation of minimal length is contradicted.

3 Only 2, 3, 4, 5, 6 are allowable by qn 1. Qn 2 prohibits 5.

4 Use qn 21.25ff.

5 If $\alpha : z \mapsto e^{i\theta}z + c$ and $\beta : z \mapsto e^{i\theta}z + d$, the image of $\alpha^{-1}\beta^{-1}\alpha\beta$ in the point group is $z \mapsto e^{-i\theta}e^{-i\phi}e^{i\theta}e^{i\phi}z = z$. If $\beta : z \mapsto e^{i\phi}\bar{z} + d$, then the image of $\alpha\beta^{-1}\alpha\beta$ in the point group is $z \mapsto \overline{e^{i\theta}}e^{i\phi}e^{i\theta}e^{i\phi}z = z$. Only translations map to the identity in the point group.

6 Whatever ϱ and β may be, $\varrho\beta^{-1}\varrho\beta$ must be a direct isometry. If it is a rotation, then W contains a translation by qn 5. The square of a glide-reflection is a translation.

7 Under the group $\langle \tau, \sigma \rangle$, the orbit of each point is a parallelogram lattice. If t is a translation in W but not in $\langle \tau, \sigma \rangle$, the point Ot lies in a parallelogram with vertices $O\tau^m\sigma^n$, $O\tau^{m+1}\sigma^n$, $O\tau^{m+1}\sigma^{n+1}$ and $O\tau^m\sigma^{n+1}$ for some integers m and n. So $O(t\tau^{-m}\sigma^{-n})$ lies in the parallelogram O, $O\tau$, $O\tau\sigma$, $O\sigma$. But it cannot lie on one of the edges of the parallelogram without contradicting the minimality of τ or σ. If it lies in the triangle $O\tau$, $O\tau\sigma$, $O\sigma$, then $O\tau\sigma(t\tau^{-m}\sigma^{-n})^{-1}$ lies in the triangle O, $O\tau$, $O\sigma$ and we then have a translation of length less than σ in $W - \langle \tau \rangle$. Since τ and σ are in different directions, $\tau^i\sigma^j = \tau^m\sigma^n \Rightarrow \tau^{i-m} = \sigma^{n-j} = 1$, so $i = m$ and $j = n$. Thus the conditions of qn 10.11 are satisfied by the subgroups $\langle \tau \rangle$ and $\langle \sigma \rangle$.

8 With point group C_2, the translation group has two cosets, namely, T and $T\alpha$. If all the translations were in one direction, the group would fix a line. With translations in two directions, no point or line is fixed, and the group is discrete by definition.

9 Four.

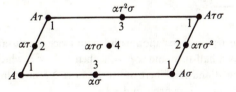

10 The translation group T has three cosets in W, T, $T\alpha$ and $T\alpha^2$. Since $\alpha^{-1}\tau\alpha$ is a translation of the same length as τ, $T = \langle \tau, \alpha^{-1}\tau\alpha \rangle$.

11 The point $A\beta$ is the centre of $\beta^{-1}\alpha\beta$. The length of the translation $\alpha\beta^{-1}$ is $\sqrt{3}$ times the distance between the centres. If τ is a translation of minimal length, this implies that the centres marked are as closely packed as possible, so all 3-centres within the hexagon have been identified. Thus the orbits of 3-centres are just the orbits under the translation group.

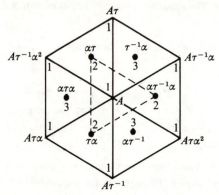

12 The translation group T has four cosets in W, namely T, $T\alpha$, $T\alpha^2$ and $T\alpha^3$. The group element $\alpha^{-1}\tau\alpha$ is a translation of the same length as τ in $W - \langle\tau\rangle$.

13 If α has centre A, then $\beta^{-1}\alpha\beta$ has centre $A\beta$. The length of the translation $\alpha\beta^{-1}$ is $\sqrt{2}$ times the distance between the centres. So the 4-centres marked and numbered in the square are as closely packed as possible, and all the 4-centres in the square have been identified. The orbits of 4-centres are just the orbits under the translation group.

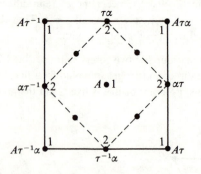

14 The product of two half-turns is a translation with length twice the distance between the centres. If β is a half-turn and $\alpha^2\beta = \tau$, then $\beta = \alpha^2\tau$, which has centre midway between A and $A\tau$. There are three more 2-centres in the square of qn 13, belonging to one orbit under $\langle\alpha\rangle$.

15 The translation group T has six cosets of the form $T\alpha^i$. The group element $\alpha^{-1}\tau\alpha$ is a translation of the same length as τ in $W - \langle\tau\rangle$.

16 The group element $\alpha\beta^{-1}\alpha^{-1}\beta$ is a translation as in qn 5. The point $A\eta$ is the centre of $\eta\alpha\eta$.

17 The group element $\beta\alpha$ is a rotation through $\frac{2}{3}\pi$ with centre at the centroid of the triangle A, B, $B\alpha$. $A\beta\alpha = B$, so the 6-centres lie in one orbit. There are six 3-centres in one orbit of $\langle\alpha\rangle$ inside the hexagon with vertices $B\alpha^i$. From qn 11 this number is maximal.

18 The axis of $z \mapsto e^{i\theta}\bar{z}$ is $\{re^{\frac{1}{2}i\theta}|r \in \mathbb{R}\}$. If $\gamma: z \mapsto e^{i\theta}\bar{z} + c$, then $\gamma^2: z \mapsto z + e^{i\theta}\bar{c} + c = z + (e^{\frac{1}{2}i\theta}\bar{c} + e^{-\frac{1}{2}i\theta}c)e^{\frac{1}{2}i\theta}$ and $e^{\frac{1}{2}i\theta}\bar{c} + e^{-\frac{1}{2}i\theta}c$ is real. If $e^{i\theta}\bar{c} + c = 0$, the result follows from qn 3.26.

19 Because D_1 has order 2, the subgroup of direct isometries has exactly two cosets. The line l would be fixed by W if all translations were parallel to l. From qn 3.30 $\beta\varrho = \varrho\beta$, and because α and β are translations, $\alpha\beta = \beta\alpha$, so $\alpha^{-1}\gamma^{-1}\alpha\gamma = \alpha^{-1}\varrho^{-1}\alpha\varrho$. Now $\alpha^{-1}\varrho\alpha$ is the reflection in $l\alpha$ which is parallel to l, so $\alpha^{-1}\gamma^{-1}\alpha\gamma$ is a translation in W in a direction perpendicular to l. If γ were a glide-reflection, then γ^2 would be a translation parallel to l. If all the translations in W were perpendicular to l, then every line perpendicular to l would be fixed by W. $(\gamma\mu)^2 = (\gamma\mu\gamma)\mu$, which is a product of translations. If $A \in l$, then l is the perpendicular bisector of $A\mu$ and $A\gamma\mu\gamma$. Also, $AA\mu$ is not perpendicular to l and so $(\gamma\mu\gamma)\mu$ is not the identity.

20 If γ is a reflection, $\gamma^2 = 1$. If γ is a glide-reflection, γ^2 is a translation in the same direction as τ. But τ is a translation of minimal length in this direction, so the discreteness of W implies that γ^2 is equal to some power of τ. If m is even, then $\gamma\tau^{-\frac{1}{2}m}$ is an opposite isometry of order 2. If m is odd, let $\delta = \gamma\tau^{-\frac{1}{2}(m-1)}$, then $\delta^2 = \tau$ and $\delta^m = \gamma$, so $\langle\delta\rangle = \langle\gamma, \tau\rangle$.

21 Because the point group is D_1, all the reflections in W have parallel axes. If ϱ_1 and ϱ_2 are reflections in parallel axes, $\varrho_1\varrho_2$ is a translation in a direction perpendicular to their axes and through twice the distance between them. If ϱ_1 and ϱ_2 are in W with axes as close together as possible, then every other reflection axis lies in the same orbit as one of these two, under the translation group.

22 Since $\sigma\delta\sigma = \delta$, every element of $\langle\delta, \sigma\rangle$ has the form $\delta^i\sigma^j$. A reflection would be an element of order 2, but $(\delta^i\sigma^j)^2 = \delta^i\sigma^j\delta^i\sigma^j = \delta^{2i} \neq 1$ unless $i = 0$.

23 The isometry $\mu^{-1}\delta$ fixes points midway between l and $l\mu$.

24 Let O be a point on the axis of the reflection ϱ. Consider the fundamental parallelogram of the translation group O, $O\mu$, $O\tau$, $O\mu\varrho$, with $OO\tau = l$. The lines l, $l\mu$ and $l\mu\varrho$ are axes of reflections which intersect the parallelogram. If there were any more such, the product of two reflections in W would give a translation in the direction of σ, but of shorter length. So the reflection axes lie in one orbit under the translation group. Now the opposite isometries $\mu\varrho$ and $\varrho\mu$, move the midpoints of $OO\mu\varrho$ and $OO\mu$ respectively, parallel to l, and since these points do not lie on the axes of reflections, these lines parallel to l are the axes of glide-reflections. Moreover, $\mu^{-1}(\mu\varrho)\mu = \varrho\mu$, so these glide-reflection axes lie in the same orbit of the translation group. Every glide-reflection in W has the form $\varrho\tau'$ for some translation τ' in W and the axis of the glide-reflection $\varrho\tau'$ passes through the midpoint of $OO\varrho\tau' = OO\tau'$, so there are no more glide-reflection axes intersecting the fundamental parallelogram.

25 The images of γ and δ in the point group are reflections in perpendicular axes, so the image of their product is a half-turn and $\gamma\delta$ is a half-turn. Now D_2 has order 4, so the translation group T has four cosets in W, namely, T, $T\gamma$, $T\delta$ and $T\gamma\delta$.

26 If O is the point of intersection of the axes of ϱ_1 and ϱ_2, then $\varrho_1\varrho_2\tau$, $\varrho_1\varrho_2\sigma$ and $\varrho_1\varrho_2\sigma\tau$ are half-turns with centres in the rectangle O, $O\tau$, $O\sigma$, $O\sigma\tau$. There cannot be more 2-centres in this fundamental parallelogram of the translation group because the product of two half-turns is a translation through twice the distance between the centres. The reflections $\varrho_1\sigma$ and $\varrho_2\tau$ have axes parallel to the sides and bisecting the rectangle. There cannot be more reflection axes intersecting this fundamental parallelogram because the product of two reflections in parallel axes is a translation through twice the distance between them.

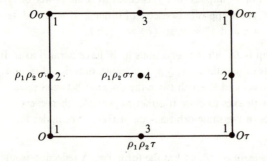

27 If α is a half-turn with centre on the axis of a glide-reflection γ, $\alpha\gamma$ is a reflection with a perpendicular axis. The direct isometries $\lambda_1\lambda_2$, $\lambda_2\lambda_1$, $\lambda_1^{-1}\lambda_2$ and $\lambda_2\lambda_1^{-1}$ are half-turns with centres within a fundamental rectangle with middle lines the axes of λ_1 and λ_2 intersecting at O. There cannot be more 2-centres in the rectangle because of the length of the minimal translations. Since $\lambda_1^{-1}(\lambda_1\lambda_2)\lambda_1 = \lambda_2\lambda_1$ and $\lambda_1(\lambda_1^{-1}\lambda_2)\lambda_1^{-1} = \lambda_2\lambda_1^{-1}$, the four 2-centres belong to at most two orbits. A detailed consideration of the four types of isometry shows that no element of W can map the centre of $\lambda_1\lambda_2$ to the centre of $\lambda_2\lambda_1^{-1}$.

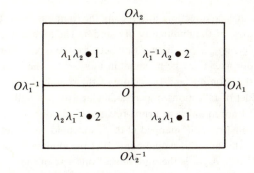

28 If a 2-centre lay on the axis of a reflection there would be reflections with axes in perpendicular directions. If α is a half-turn and ϱ is a reflection with axis not through the centre of α, then $\alpha\varrho$ is a glide-reflection with axis perpendicular to that of ϱ and passing through the centre of α. So every 2-centre lies on a glide axis. Now $\lambda\varrho$ and $\varrho\lambda$ are half-turns with centres on the axis of λ and $\lambda^{-1}(\lambda\varrho)\lambda = \varrho\lambda$ so the centres lie in the same orbit. The product of these half-turns is λ^2 which is a minimal translation in this direction. Also $\lambda\varrho\tau$ and $\varrho\lambda\tau$ are half-turns with centres on the axis of the glide-reflection $\lambda\tau$ in the fundamental rectangle $O\lambda$, $O\lambda\tau$, $O\lambda^{-1}\tau$, $O\lambda^{-1}$, where O is the point of intersection of the axis of ϱ and the axis of λ. To show that $\varrho\lambda$ and $\varrho\lambda\tau$ have centres in different orbits, consider the types of elements in W. An element that would put these centres in the same orbit would contradict the minimality of τ. This is obvious if the element were to be a translation or a half-turn. If it were to be a glide-reflection, the product with ϱ needs to be considered.

30 Let A be the point of intersection of two perpendicular axes of reflections in W. Consider the fundamental parallelogram with vertices A, $A\mu$, $A\sigma$, $A\sigma\mu^{-1}$, which is a rhombus in this case. Since there are reflections with perpendicular axes through A, the same holds for the other vertices by conjugation by the appropriate translations. Let the diagonals of this rhombus be the axes of reflections ϱ_1 and ϱ_2 and let these axes intersect at O.

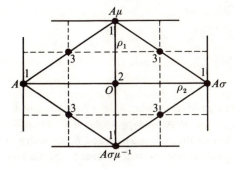

The three reflection axes in each direction, intersecting the rhombus, must be maximal in number because of the minimality of τ and σ. The axes of the glide-reflections $\mu\varrho_1$, $\mu^{-1}\varrho_1$, $\mu\varrho_2$, $\mu^{-1}\varrho_2$, lie midway between these reflection axes and, again, these four axes must be maximal in number for the same reason. The vertices of the rhombus and the centre of the rhombus are 2-centres since the product of reflections in perpendicular axes is a half-turn. The points O and A are in different orbits because if any one of the four different types of isometry in W mapped O to A, we could construct a translation in W doing this and this would contradict the minimality of σ. The centre of the half turn $\varrho_1\varrho_2\mu$ is the midpoint of one side of the rhombus. The four midpoints of the sides of the rhombus lie in one orbit under the glide-reflections $\mu\varrho_1$, $\mu^{-1}\varrho_1$, $\mu\varrho_2$, $\mu^{-1}\varrho_2$.

31 If γ is a reflection $\gamma\alpha^{-1}\gamma^2\alpha = \gamma$. If γ is a glide-reflection and A is on its axis, the two points, A and $A\gamma\alpha^{-1}\gamma^2\alpha$, are transposed by $\gamma\alpha^{-1}\gamma^2\alpha$, so qn 3.30(i) cannot hold. $\alpha^{-1}\gamma^2\alpha$ is a translation.

33 The argument and diagram of qn 11 are valid here. The existence of the reflection means that the 3-centres in orbits 2 and 3 will belong to one orbit for this group.

34 Again the argument and diagram of qn 11 hold, but the reflections here leave the three orbits of 3-centres of qn 11 distinct.

35 The direct isometries form a subgroup of index 2. The isometry $\gamma^{-1}\alpha\gamma$ is a quarter-turn with centre $A\gamma$. Every opposite isometry $\delta = \gamma\eta$ for some direct isometry η. Now $A\delta = A\gamma\eta$ and since $A\gamma$ is in orbit A, and η permutes the points of orbit A, $A\delta$ is in orbit A.

  (i) Since $A\gamma\beta^{-1} = A$, $\gamma\beta^{-1}$ is a reflection with axis through A. All the four reflections fix A, and only the four points $A\tau$, $A\tau\alpha$, $A\tau^{-1}$ and $A\tau^{-1}\alpha$ of orbit A are equidistant from A, so these are the four possible images of $A\tau$.

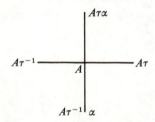

  (ii) The direct isometries are transitive on orbit B, so if $A\gamma$ is in orbit B, for some direct isometry η, $A\gamma\eta = B$. Also $A\alpha\gamma\eta = A\alpha^2\gamma\eta = A\alpha^3\gamma\eta$. Since the opposite isometries interchange the orbits A and B, the image of B lies in orbit A. But the distance from A to B is equal to the distance

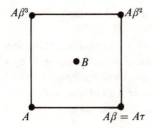

between their images under an isometry, and the only points in orbit A, distant AB from B are A, $A\beta$, $A\beta^2$ and $A\beta^3$. The point B is the midpoint of $AA\beta^2$, so γ acts on this line like a translation and must in fact be a glide-reflection with axis AB. $A(\alpha\gamma)^2 = B\alpha\gamma = A\tau$, and $B(\alpha\gamma)^2 = B\tau$, so $(\alpha\gamma)^2 = \tau$. If the glide-reflection $\gamma = \varrho\sigma = \sigma\varrho$, with ϱ a reflection and σ a translation, $\alpha^2\gamma = \alpha^2\varrho\sigma$, but $\alpha^2\varrho$ is a reflection with axis perpendicular to that of ϱ and so $(\alpha^2\varrho)\sigma$ is a reflection. Now $\alpha^{-1}(\alpha^2\gamma)\alpha$ is a reflection in W and axis perpendicular to that of $\alpha^2\gamma$. If there were reflections with axes in four directions, then two axes inclined at $\frac{1}{4}\pi$ would intersect and the point of intersection would be a 4-centre as in *p4m* (case (i)).

36 If A is a 4-centre lying on the axis of the reflection, as ϱ in qn 35(i), consider the fundamental square A, $A\tau$, $A\sigma$, $A\tau\sigma$, where $\sigma = \alpha^{-1}\tau\alpha$ and α is a quarter-turn about A. Plainly the sides of this square are all in the orbit of the axis of ϱ under $\langle\alpha, \tau\rangle$. If B is the centre of this fundamental square, $\varrho(\alpha^{-1}\tau\alpha)$ is a reflection with axis through B parallel to the sides of the

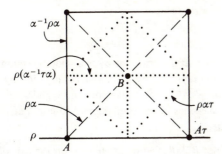

square. No points of orbit A lie on the axis of $\varrho(\alpha^{-1}\tau\alpha)$ so its orbit is distinct from that of the sides of the square. The axis of $\varrho\alpha$ is inclined at an angle of $\frac{1}{4}\pi$ to the axes of the other reflections which we have so far identified. Since the three generators of W each preserve such a direction or rotate it through $\frac{1}{2}\pi$, the axis of the reflection $\varrho\alpha$ is not in either of the two orbits of axes previously identified. The orbit of this axis includes parallel and perpendicular axes through each of the vertices of the fundamental square. We have now identified all the reflection axes because the product of reflections in parallel axes is a translation through twice the distance between them, and τ is minimal.

The transformation $\varrho\alpha\tau$ is a glide-reflection with axis parallel to the diagonal through A, passing through the midpoint of $AA\tau$. Conjugation by α gives another glide-reflection with axis through the midpoint of $AA\tau$ and then conjugation by translations gives two more axes of glide-reflections intersecting the unit square. So these axes all lie in one orbit.

37 Let A be the centre of α with the axes of the glide-reflection γ passing through A. Since $(\alpha\gamma)^2$ is a minimal translation (with the conventions of qn 35(ii)), we consider a fundamental square of which A, $A\gamma^2$ and $A(\alpha\gamma)^2$ are three of the vertices. The transformation $\alpha^2\gamma$ and its conjugates by α, γ and $\gamma\alpha$ give four reflections with axes parallel to the diagonals of the fundamental square. Six glide-reflections are obtained from γ by repeated conjugation by α and γ with axes parallel to the diagonals of the fundamental square and intersecting the square.

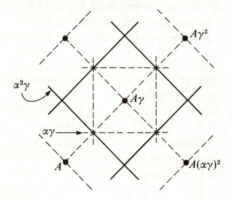

These axes all contain 4-centres. The transformation $\alpha\gamma$ is a glide-reflection with axis parallel to the translation $(\alpha\gamma)^2$ and by successive conjugation by the quarter-turns α^{-1}, $(\alpha\gamma^2)^{-1}$ and $\gamma^{-2}\alpha^{-1}\gamma^2$ we obtain four more glide-reflections with axes parallel to the sides of the fundamental square, but not passing through any 4-centre.

38 For any $\beta \in W$, $A\beta$ is the centre of $\beta^{-1}\alpha\beta$. The point $A\varrho\alpha$ lies on the axis of the reflection ϱ because α is a rotation through $\frac{1}{2}\pi$. The reflection $\varrho\alpha$ has axis $AA\tau$. $\tau\alpha\tau\alpha^{-1} = \sigma$.

39 The 6-centres lie in one orbit and the 3-centres lie in one orbit from qn 17. If α is a rotation through $\frac{1}{3}\pi$ about A and β is a rotation through $\frac{1}{3}\pi$ about B then $\beta\alpha\beta$ is a half-turn about the midpoint of AB. If A and B are 6-centres the minimal distance apart (the length of a minimal translation) and A' and B' are also 6-centres the minimal distance apart, then there is an isometry in the group, say δ, such that $A\delta = A'$. Then for some i,

$A\alpha^i\delta = A'$ and $B\alpha^i\delta = B'$, so that the midpoint of AB is mapped to the midpoint of $A'B'$ by δ. Thus the 2-centres form one orbit. If the reflections ϱ, $\varrho\alpha^2 = \alpha^{-1}\varrho\alpha$ and $\varrho\alpha^4 = \alpha^{-2}\varrho\alpha^2$ have axes parallel to a minimal translation, then, $\varrho\alpha$, $\varrho\alpha^3 = \alpha^{-1}(\varrho\alpha)\alpha$ and $\varrho\alpha^5 = \alpha^{-2}(\varrho\alpha)\alpha^2$ do not have axes in such a direction and their axes belong to a different orbit.

40 Both in this equation and in the one that follows we suppose that we have two groups of the same type to prove isomorphic. A suffix 1 will denote an element of the first group, and a suffix 2 will denote an element of the second.

p1

Take the same origin **O**. Then $\mathbf{O}\tau_1$ and $\mathbf{O}\sigma_1$ are a pair of basis vectors and $\mathbf{O}\tau_2$ and $\mathbf{O}\sigma_2$ are also a pair of basis vectors. There exists a nonsingular linear transformation $\mu : \mathbf{v} \mapsto \mathbf{v}M$ such that $(\mathbf{O}\tau_1)\mu = \mathbf{O}\tau_2$ and $(\mathbf{O}\sigma_1)\mu = \mathbf{O}\sigma_2$. Then $\mu^{-1}\tau_1\mu = \tau_2$ and $\mu^{-1}\sigma_1\mu = \sigma_2$, so the groups are isomorphic under an inner automorphism of the affine group.

p2

Take origins $\mathbf{O}_1$ and $\mathbf{O}_2$ at 2-centres. If $\mathbf{O}_2$ has the position vector **c** relative to $\mathbf{O}_1$, and the nonsingular transformation $\mathbf{v} \mapsto \mathbf{v}M$ fixing the origin $\mathbf{O}_1$ maps $\mathbf{O}_1\tau_1$ to $\mathbf{O}_2\tau_2 - \mathbf{c}$ and $\mathbf{O}_1\sigma_1$ to $\mathbf{O}_2\sigma_2 - \mathbf{c}$, and $\mu : \mathbf{v} \mapsto \mathbf{v}M + \mathbf{c}$, then $\mu^{-1}\tau_1\mu = \tau_2$, $\mu^{-1}\sigma_1\mu = \sigma_2$ and $\mu^{-1}\alpha_1\mu = \alpha_2$ where α_1 and α_2 are the half-turns with centres $\mathbf{O}_1$ and $\mathbf{O}_2$ respectively. So the groups are isomorphic under an inner automorphism of the affine group. The methods indicated here apply also to the types *pm*, *pg*, *cm*, *pmm*, *pmg*, *pgg* and *cmm*.

41 *p3*.

Take origins $\mathbf{O}_1$ and $\mathbf{O}_2$ at 3-centres. if $\mathbf{O}_2$ has the position vector **c** relative to $\mathbf{O}_1$, and the similarity $\mathbf{v} \mapsto \mathbf{v}M$ fixes the origin $\mathbf{O}_1$ and maps $\mathbf{O}_1\tau_1$ to $\mathbf{O}_2\tau_2 - \mathbf{c}$, then if $\mu : \mathbf{v} \mapsto \mathbf{v}M + \mathbf{c}$, we have $\mu^{-1}\tau_1\mu = \tau_2$ and $\mu^{-1}\alpha_1\mu = \alpha_2$ where α_1 and α_2 are rotations through $\frac{2}{3}\pi$ with centres $\mathbf{O}_1$ and $\mathbf{O}_2$ respectively. This method can be applied also to the types of group *p4*, *p6*, *p31m*, *p3m1*, *p4m*, *p4g* and *p6m*. The use of a similarity in these cases (not just an affine transformation) is necessary to preserve angles.

Bibliography

Bartlow, T. L., 1972, 'An historical note on the parity of permutations', *Amer. Math. Monthly*, **79**, 766–9.

Birkhoff, G. and MacLane, S., 1953. *A Survey of Modern Algebra*, Macmillan, New York.

Bourbaki, N., 1960, *Eléments d'histoire des mathématiques*, Hermann, Paris.

Budden, F. J., 1978, *The Fascination of Groups*, Cambridge University Press.

Burn, R. P., 1973, 'Geometrical illustrations of group theoretical concepts', *Math. Gaz.*, **57**, 110–19.

Burn, R. P., 1977, Groups of linear transformations, *Math. Gaz.*, **61**, 273–9.

Cajori, F., 1952, *A History of Mathematical Notations* II, Open Court.

Cajori, F., 1961, *A History of Mathematics*, Macmillan, New York.

Carmichael, R. D., 1956, *Introduction to the Theory of Groups of Finite Order*, Dover.

Cayley, A., 1854, 'On the theory of groups, as depending on the symbolic equation $\theta^n = 1$, *Phil. Mag.*, **7**, 40–7, *Coll. Math. Works*, **II**, 123–30.

Cayley, A., 1858, A memoir on the theory of matrices, *Phil. Trans. R.S.*, **148**, *Coll. Math. Works*, **II**, 475–96.

Cayley, A., 1878, 'The theory of groups', *Amer. J. Math.*, **1**, 50–2, *Coll. Math. Works*, **X**, 401–3.

Coxeter, H. S. M., 1969, *Introduction to Geometry*, Wiley.

Coxeter, H. S. M., 1973, *Regular Polytopes*, Dover.

Coexter, H. S. M., 1974, *Regular Complex Polytopes*, Cambridge University Press.

Curtis, M. L., 1979, *Matrix Groups*, Springer-Verlag.

Dieudonné, J., 1962, *La Géometrie des Groupes Classiques*, Springer-Verlag.

Dieudonné, J., 1969, *Linear Algebra and Geometry*, Kershaw.

Ford, L. R., 1951, *Automorphic Functions*, Chelsea.

Forder, H. G., 1960, *Geometry*, Hutchinson.

Fraleigh, J. B., 1976, *A First Course in Abstract Algebra*, Addison-Wesley.

Fricke, R. and Klein, F., 1897, *Vorlesungen über die Theorie der automorphen Functionen* I, Teubner.

Gardiner, C. F., 1980, *A First Course in Group Theory*, Springer-Verlag.

Gauss, C. F., 1801, *Disquisitiones Arithmeticae*, tr. English 1965, Yale.

Green, J. A., 1965, *Sets and Groups*, Routledge and Kegan Paul.

Hilbert, D. and Cohn-Vossen, S., 1952, *Geometry and the Imagination*, Chelsea.

Hille, E., 1959, *Analytic Function Theory* I, Blaisdell.

Jordan, C., 1870, *Traité des Substitutions*, Gauthier-Villars (reprinted 1957).

Klein, F., 1884, *Lectures on the Icosahedron*, tr. English 1956, Dover.

Knopp, K., 1952, *Elements of the Theory of Functions*, Dover.

Ledermann, W., 1960, *Complex Numbers*, Routledge and Kegan Paul.

Lockwood, E. H. and Macmillan, R. H., 1978, *Geometric Symmetry*, Cambridge University Press.

Martin, G. E., 1982, *Transformation Geometry*, Springer-Verlag.

Maxwell, E. A., 1965, *Algebraic Structure and Matrices* II, Cambridge University Press.

Miller, G. A., 1935, 'History of the theory of groups to 1900', *Coll. Works*, vol. 1, pp. 427–67, University of Illinois, Urbana.

Netto, E., 1880, *Theory of Substitutions*, tr. English 1892, Chelsea.

Neumann, P. M., Stoy, G. A. and Thompson, E. C., 1980, *Groups and Geometry*, Mathematical Institute, Oxford.

Nový, A., 1973, *Origins of Modern Algebra*, Noordhoff International.

Pedoe, D., 1970, *A Course of Geometry for Colleges and Universities*, Cambridge University Press.

Pólya, G., 1924, 'Über die Analogie der Kristallsymmetrie in der Ebene', *Zeitschrift für Kristallographie*, **60**, 278–82.

Rees, E. G., 1983, *Notes on Geometry*, Springer-Verlag.

Rotman, J. J., 1973, *The Theory of Groups*, Allyn and Bacon.

Schattschneider, D., 1978, 'The plane symmetry groups', *Amer. Math. Monthly*, **85**, 439–50.

School Mathematics Project, 1970, *Additional Mathematics Part 1*, Cambridge University Press.

Schwerdtfeger, H., 1979, *Geometry of Complex Numbers*, Dover.

Speiser, A., 1927, *Theorie der Gruppen von endliche Ordnung*, Springer-Verlag.

Steinhaus, H. 1960, *Mathematical Snapshots*, Oxford University Press.

Weyl, H., 1952, *Symmetry*, Princeton.

Yaglom, I. M., 1973, *Geometric Transformations* III, Math. Assn. America.

Index

References refer to questions, e.g. 2.32 means chapter 2, question 32